Artificial Intelligence and Machine Learning Techniques in Engineering and Management

Komaragiri Srinivasa Raju · Dasika Nagesh Kumar

Artificial Intelligence and Machine Learning Techniques in Engineering and Management

Springer

Komaragiri Srinivasa Raju
Department of Civil Engineering
BITS Pilani, Hyderabad Campus
Hyderabad, Telangana, India

Dasika Nagesh Kumar
Department of Civil Engineering
Indian Institute of Science Bangalore
Bengaluru, Karnataka, India

Lyles School of Civil and Construction Engineering
Purdue University
West Lafayette, IN, USA

ISBN 978-981-96-2623-6 ISBN 978-981-96-2621-2 (eBook)
https://doi.org/10.1007/978-981-96-2621-2

This Springer imprint is published by the registered company Springer Nature Singapore Pte Ltd.
The registered company address is: 152 Beach Road, #21-01/04 Gateway East, Singapore 189721, Singapore

Foreword

The 2024 Nobel Prize in Physics was awarded to John J. Hopfield and Geoffrey Hinton *for foundational discoveries and inventions that enable machine learning with artificial neural networks.* Two of the three winners of the 2024 Nobel Prize in Chemistry, Demis Hassabis and John Jumper, both from Google Deepmind, were cited for developing an AI model to solve a 50-year-old problem: predicting complex structures of proteins. These coveted prizes to Artificial Intelligence (AI) researchers bear testimony to the sweeping influence of AI.

Driven by extraordinary strides made over the years, especially in the past decade, AI now has the potential to fundamentally transform human civilization. Its importance is now recognized by societies across the globe as a key technology with the ability to solve some of the most complex societal and engineering problems of our times such as universal access to healthcare and education, efficient transportation, increased efficiency in providing e-governance services to the public, etc. To harness the power of AI, large-scale national and international efforts are underway.

AI has now matured to a level where AI applications are beginning to impact our daily lives: generative AI tools like ChatGPT and Gemini are now extensively used by researchers, students, and even public. Among the myriad of disciplines impacted by AI, engineering and management disciplines occupy a prominent position. AI and data science are now providing a major tool box to solve a wide spectrum of problems in engineering and management.

There is a large corpus of textbooks and research monographs on the foundations, theory, and advances in artificial intelligence, machine learning (ML), and deep learning (DL). There is, however, an urgent need for a book that provides a convenient, friendly, and yet rigorous treatment of AI and ML techniques to researchers, professionals, and students engaged in core engineering disciplines and also core management topics. This gap is splendidly filled by authors Srinivasa Raju and Nagesh Kumar, by bringing out their fine and timely textbook *Artificial Intelligence and Machine Learning Techniques in Engineering and Management.*

This is a nice book accessible to anyone seeking to clearly understand and rigorously apply AI techniques to problems in engineering and management disciplines. In particular, it will be a precious resource to undergraduate, master's, and doctoral students applying AI and data science to their projects and research problems. The coverage of topics in machine learning and deep learning models is gentle and thorough, focussing on the main principles. The illustrative numerical examples, completely worked out, elevate the utility and understandability of the contents. The final chapter is especially valuable, with more than 200 case studies reviewed; this will be a goldmine to look up detailed studies of real-world problems.

The authors must be congratulated for conceptualizing a much needed AIML companion to students and researchers and for presenting the content in a lucid manner. For engineering and management audience, this book is a lovely resource on a live and lively subject.

Y. Narahari
Honorary Professor
Department of Computer Science
and Automation
Indian Institute of Science Bangalore
Bengaluru, Karnataka, India
https://gtl.csa.iisc.ac.in/hari/

Preface

Artificial Intelligence (AI) is becoming familiar due to the minimum requirement of data, facilitating accurate predictions, and minimal necessity of understanding the physics behind input–output relationships. Its potential to tackle non-linear and complex problems with greater flexibility is an added advantage. Its applications in engineering, management, and allied fields are growing exponentially. Over time, numerous experts introduced books and developed blogs on the theme, which are primarily theoretical. However, the proposed book amalgamates relevant theory, numerical problems, case studies, and recent advances wherever possible. We believe that this new dimension will greatly benefit present-generation researchers and students.

The present book consists of seven chapters: (1) an introduction; (2) a description of performance indicators; (3) classical machine learning algorithms; (4) advanced machine learning algorithms; (5) fuzzy logic-based modelling algorithms; (6) emerging research areas, topics including, Blockchain, recent ML techniques, evolutionary algorithms, AI tools, the Internet of Things, big data, decision support systems, Taguchi design of experiments, data augmentation, and cross-validation; (7) representative case studies. The appendix covers representative AI tools, data sources related to AI, books, and journals on AI. The present book can support undergraduate, postgraduate, and Ph.D. students in AI, Data Mining, and Soft Computing in Engineering and Management and allied fields.

We are grateful to Prof. Yadati Narahari, Department of Computer Science and Automation, Indian Institute of Science, Bengaluru, who consented to write a Foreword for the book.

Special acknowledgments to Vogeti Rishith Kumar for posing thought-provoking, out-of-box questions, providing lots of input, and unstinting support wherever necessary. Heartfelt gratitude to Sistla Shashank, Prof. Alivelu Manga Parimi, Deepjyoti Deb, Dr. R. Madhuri, R. Bhavi Tej, Dr. Sriman Pankaj, Kathan Pranav Naik, Y. Sai Kiran, Pratyush Pandey, P. Sagar Subhash, Bhavesh Rahul Mishra, Harshal Nayan Rathi, Rahul Jauhari, Rishabh Daga, Ayushman Kar, Aakash Bansal, Kaustav Chatterji, and L. Ashoka Vardhan (who are presently or formerly associated with BITS) who contributed immensely for the book. Also, thanks to Prof. M. Janga Reddy (IIT Bombay), Prof. Shishir Gaur (IIT BHU), Prof. D. Graillot (EMSE France), Prof. D. V. Morankar (College of Military Engineering, Pune), and many others who supported us from time to time.

We referred to a number of research papers and many blog sites related to AI. Overall, they shaped the book in its present form. We acknowledge LINDO SYSTEMS INC. for providing access to the LINGO software trial version, Scopus for research data analysis, and Python for programming support.

We have incorporated a few portions from some of our published research papers, either utilizing CC BY 4.0 and CC BY-NC-ND 4.0 licenses under the open access category or taking permissions in case of non-open access category journals. All these research papers were referred at suitable places. We wholeheartedly express gratefulness to the publishers of these journals, IWA, Springer, ASCE and Wiley. We extend thanks to all the co-authors of the papers for their constant encouragement and support in realizing our plan to publish this book.

We made the best possible efforts to quote all the sources in the form of acknowledgements or references, but still, some would have been missed. We will incorporate them upon notice in the upcoming editions.

Professor Raju appreciates the institute leadership for providing the necessary ecosystem for writing this book. He acknowledges the help of his wife, Gayathri Devi; Daughter, Sai Swetha; and son, Sai Satvik; and Parents, Gopala Rao and Varalakshmi, for their unstinting support. He thanks Prof. A. Vasan, Subbulakshmi Vasan, Dr. K. Nagajyothi, and Mr. B. Surendra for their motivating support. Professor Nagesh acknowledges the support of his wife Padma, daughter Sruthi, son Saketh, and parents Subrahmanyam and Lakshmi.

We sincerely thank Sri D. V. Subrahmanyam for diligently checking the manuscript and proofs.

We wish to thank all our colleagues, friends, and students who encouraged us from time to time with pleasant inquiries and inputs, which undoubtedly accelerated the writing of the book.

Lastly, we are thankful to Swati Meherishi and her team at Springer for processing the book in a timely manner.

Komaragiri Srinivasa Raju
Senior Professor
Department of Civil Engineering
BITS Pilani, Hyderabad Campus
Hyderabad, Telangana, India
ksraju@hyderabad.bits-pilani.ac.in
https://www.bits-pilani.ac.in/hyderabad/ksraju/Profile

Dasika Nagesh Kumar, FNA, FASc, FNASc
Professor
Department of Civil Engineering
Associate Faculty, Divecha Centre for Climate Change
Associate Faculty, Centre for Earth Sciences
Indian Institute of Science Bangalore
Bengaluru, Karnataka, India
nagesh@iisc.ac.in
http://www.civil.iisc.ernet.in/~nagesh

Edward M. Curtis Visiting Professor
Lyles School of Civil and Construction Engineering
Purdue University
West Lafayette, IN, USA

Contents

About the Authors

Dr. Komaragiri Srinivasa Raju is a Senior Professor at the Department of Civil Engineering, BITS Pilani—Hyderabad Campus, India. He completed Ph.D. from the Indian Institute of Technology Kharagpur. His main research interests include artificial intelligence, machine learning, the impact of climate change, and multi-criterion decision-making. He has authored three books along with Prof. Dasika Nagesh Kumar, IISc, Bengaluru, namely, *Fluid Mechanics: Problem-Solving Using MATLAB* (2020); *Impact of Climate Change on Water Resources* (2018); and *Multi-criterion Analysis in Engineering and Management* (2010). He published more than 58 journal papers, 12 book chapters, and 88 conference papers. He edited five books and was a guest editor for three special issues of journals. He has been a reviewer for more than 45 international journals. Presently, he is Managing Editor of the *Journal of Water and Climate Change* and Associate Editor of *ISH Journal of Hydraulic Engineering*. He is also an Editorial Board Member of Scientific Reports-Springer Nature.

Dr. Dasika Nagesh Kumar is Professor at the Department of Civil Engineering at the Indian Institute of Science (IISc), Bengaluru, India. He completed Ph.D. from IISc Bangalore. He is presently working as Edward M. Curtis Visiting Professor, Lyles School of Civil and Construction Engineering, Purdue University, West Lafayette, USA, on sabbatical from IISc. He is Fellow of the Indian Academy of Sciences, Indian National Science Academy, and National Academy of Sciences. He held the Prof. Satish Dhawan Chair Professor position from 2018 to 2021. He was the Chairman, Centre for Earth Sciences, IISc, during 2014–2020. He was a Boyscast Fellow with Utah Water Research Laboratory, Utah State University, Logan, UT, USA, in 1999 and Visiting Professor in EMSE, St. Etienne, France in 2012. His research interests include climate hydrology, climate change, water resource systems, deep learning, evolutionary algorithms, fuzzy logic, MCDM, and Remote sensing and GIS applications in water resource engineering. He has supervised 10 Post-docs and 22 Ph.Ds. He is co-author of eight books and published more than 240 papers. He has received funding support of more than INR 50 crores for sponsored research. He is Editor-in-chief of *Journal of Water and Climate Change* and Associate Editor of *ASCE Journal of Hydrologic Engineering.*

Acronyms

AARLF	Average Absolute Relative Loss Function
ACO	Ant Colony Optimization
AdaBoost	Adaptive Boosting
AI	Artificial Intelligence
ALO	Ant Lion Optimizer
ANFIS	Adaptive Neuro-Fuzzy Inference System
ANN	Artificial Neural Networks
ANOVA	Analysis of Variance
AUC-ROC	Area Under the Curve-Receiver Operating Characteristic
Bi-LSTM	Bi-directional Long Short-Term Memory
BP	Back-Propagation
BRT	Boosted Regression Tree
CART	Classification And Regression Tree
CatBoost	Categorical Boosting
ChatGPT	Chat Generative Pre-trained Transformer
CNN	Convolutional Neural Networks
CNN-LSTM	Convolutional Neural Networks-Long Short-Term Memory
CS	Constraint Space
CSA	Cuckoo Search Algorithm
DE	Differential Evolution
DHL	Differential Hebbian Learning
DL	Deep Learning
DNN	Deep Neural Networks
DSS	Decision Support Systems
DT	Decision Tree
DV	Decision Variables
EA	Evolutionary Algorithms
ELM	Extreme Learning Machine
FCM	Fuzzy Cognitive Mapping
FCMe (or FCA)	Fuzzy Cluster Means (or Fuzzy Cluster Analysis)
FFBP	Feed-Forward with Back-Propagation

FIS	Fuzzy Inference Systems
FL	Federated Learning
FN	False Negative
FNR	False Negative Rate
FP	False Positive
FPR	False Positive Rate
FSS	Fractional Skill Score
GA	Genetic Algorithm
GB	Gradient Boosting
GBDT	Gradient Boosting Decision Tree
GRNN	General Regression Neural Network
GRU	Gated Recurrent Unit
GWO	Grey Wolf Optimizer
HC	Hierarchical Clustering
HGSO	Henry Gas Solubility Optimization
HL	Hebbian Learning
ICA	Imperialist Competitive Algorithm
IoT	Internet of Things
KGE	Kling Gupta Efficiency
KMe	K-Means clustering
KNN	K-Nearest Neighbour
LFM	LSTM-RF Framework with Multitasking
LFS	LSTM-RF Framework with Single tasking
LGBoost	Light Gradient Boosting
LiR	Linear Regression
LP	Linear Programming
LR	Logistic Regression
LSTM	Long Short-Term Memory
MARS	Multivariate Adaptive Regression Spline
ML	Machine Learning
MLiR	Multiple Linear Regression
MLP	Multilayer Perceptron
MSLF	Mean Square Loss Function
NAS	Neural Architecture Search
NB	Naïve Bayes
NGBoost	Natural Gradient Boosting
NHL	Non-linear Hebbian Learning
NLP	Natural Language Processing
Non-LP	Non-linear Programming
NRMSLF	Normalized Root Mean Square Loss Function
NSE	Nash–Sutcliffe Efficiency
NSGA-II	Non-dominated Sorting Genetic Algorithm-II
NSLF	Normalized Standard Loss Function
OF	Objective Function
PCA	Principal Component Analysis

PoET	Proof of Elapsed Time
PoS	Proof of Stake
PoW	Proof of Work
PSO	Particle Swarm Optimization
QP	Quadratic Programming
R_0	The highest possible R based on the perception of the user
R (or CC)	Correlation Coefficient
RBF	Radial Basis Function
ReLU	Rectified Linear Unit
RF	Random Forest
RMSLF	Root Mean Square Loss Function
RNN	Recurrent Neural Networks
RSA	Reptile Search Algorithm
RSM	Response Surface Methodology
SA	Simulated Annealing
SC	Subtractive Clustering
SHAP	SHapley Additive exPlanations
SI	Swarm Intelligence
SS	Search (Decision) Space
SSLF	Sum of Square Loss Function
SVM	Support Vector Machine
SVR	Support Vector Regression
TEE	Trusted Execution Environment
TLBO	Teaching Learning-Based Optimization
TN	True Negative
TNR	True Negative Rate
TP	True Positive
TPR	True Positive Rate
TSS	Taylor Skill Score
WGAN	Wasserstein Generative Adversarial Networks
WI	Willmott Index
WNN	Wavelet Neural Networks
XGBoost	EXtreme Gradient Boosting

List of Figures

List of Tables

Chapter 1
Introduction

1.1 Introduction

Artificial intelligence (AI) is a comprehensive multidisciplinary research area that can mimic human intellect as effectively as achievable. Some tasks that are expected to be simulated are learning, reasoning, perceiving, recognizing patterns, and decision-making. This process is also likely to minimize hindrances based on previous experiences. There are two primary classifications of AI based on functionality, which are as follows:

- Strong AI: Machines that can understand and analyze problems in various domains like humans. However, it has not yet reached complete reality. It is also termed as general AI.
- Weak AI: Works on a specific activity or function. If data is related to the heart, the experience will work for that particular task effectively, not for other domains. It is also termed as narrow AI.

A number of researchers viewed AI as a supporting mechanism for automation (Nilsson, 2009). However, most of the time, human intervention is necessary to understand the outcomes of AI for possible implementation with minimal challenges.

Two important sub-categories of AI are (a) Machine Learning (ML) and (b) Deep Learning (DL). These are employed as vehicles to accomplish AI (Fig. 1.1).

A brief description of ML and DL is as follows (Russell & Norvig, 2010): ML empowers machines to learn from available datasets without explicit programming and architecture. It uses data trends and statistical inferences to predict. DL is a sub-category in ML that utilizes the philosophy of multiple hidden layers to capture complex data phenomena automatically. The critical differences between ML and DL are the feature extraction process, data requirement, and computational resources (Alaskar & Saba, 2021; Janiesch et al., 2021).

K. Srinivasa Raju and D. Nagesh Kumar, *Artificial Intelligence and Machine Learning Techniques in Engineering and Management*,
https://doi.org/10.1007/978-981-96-2621-2_1

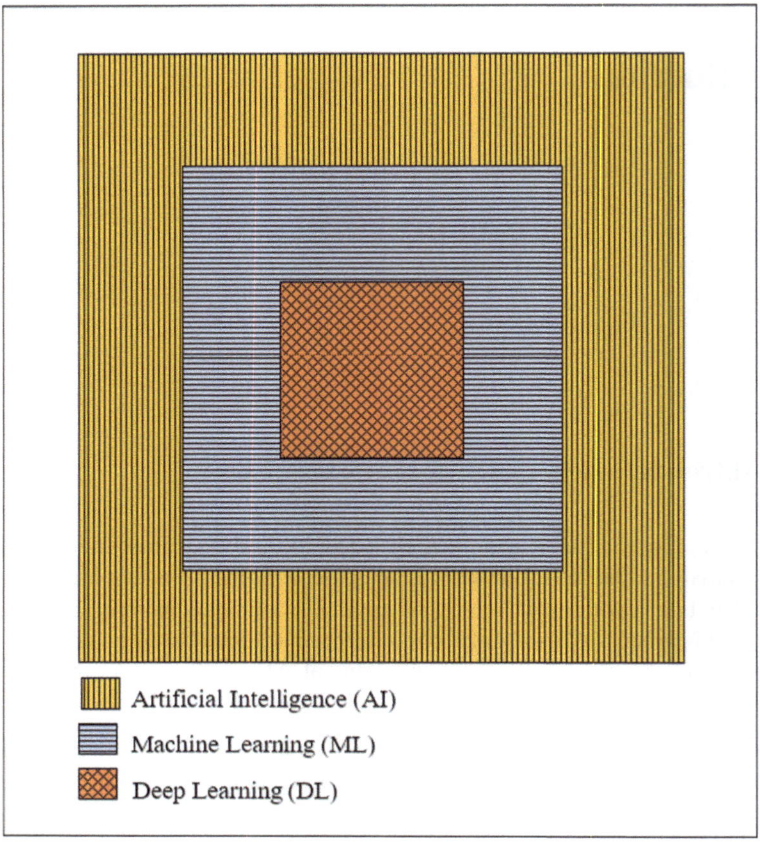

Fig. 1.1 Classification of AI into ML and DL

- DL does not require a feature extraction process. It extracts information as part of the learning, making it more efficient in prediction. On the contrary, ML requires an extensive extraction process to improve the model's performance.
- DL can work exceptionally well even with high-dimensional and unstructured raw data situations. On the other hand, ML requires pre-processing of data to ensure high performance.
- Interpretability of the prediction process is relatively more complex in DL than in ML.
- DL requires high computational resources due to its complex mathematical intricacies and hardware. In the case of ML, the requirements for computational resources are less than those for DL.

Some of the algorithms falling under ML and DL are presented (in alphabetical order) below for the reader's benefit (a detailed discussion of some of these algorithms is presented in Chaps. 3–5). In addition, related terminology is presented here for the reader to understand effectively.

- Adaptive Boosting (AdaBoost) refers to ensemble methods that make a strong learner from several weak learners. It concentrates on situations with high-loss functions to enrich total performance.
- Adaptive Neuro-Fuzzy Inference System (ANFIS) facilitates non-linear relationships, quick learning capability, and adaptive inferences to predict complex situations reasonably. It can also handle noisy or inconsistent data effectively.
- Artificial Neural Networks (ANN) can build non-linear associations between inputs and outputs. Their architecture comprises several layers. Every individual layer has a number of layers. Here, the output from each layer contributes to the succeeding layer.
- Categorical Boosting (CatBoost) is similar to AdaBoost. In this context, a Decision Tree (DT) is established on a symmetricwise strategy considered to tackle categorical features competently.
- Convolutional Neural Networks (CNN) are developed on the perception of local neural connectivity stimulated by the cognitive structure of the animal visual cortex.
- eXtreme Gradient Boosting (XGBoost) is an ensemble technique established in levelwise (or depthwise) form. It utilizes a principle identical to that of CatBoost. However, formation is governed by the depth of the tree.
- Extreme Learning Machine (ELM) employs feed-forward networks with one hidden layer. It converges faster than several traditional algorithms and will likely reach a global optimal solution.
- Fuzzy CNN and Fuzzy LSTM (Long Short-Term Memory) capitalize on the advantages of CNN, LSTM, and fuzzy reasoning. They handle imprecise data and efficiently establish relationships.
- K-Nearest Neighbour (KNN) stores all the datasets and classifies new datasets built on distance functions related to the stored datasets.
- Light Gradient Boosting (LGBoost) uses a leafwise strategy that enables data to be facilitated quicker than conventional level-based techniques.
- Linear Regression (LiR) establishes a linear association involving independent and dependent variables.
- Logistic Regression (LR) predicts classes of a binary nature and utilizes the logistic function.
- LSTM utilizes memory blocks. These blocks function as neurons in the hidden layers with the help of sigmoid and hyperbolic tangent functions. Information from layers is traversed through gating units to obtain the output.
- Multi-adaptive Regression Splines (MARS) can establish flexible and interpretable relationships between the variables using knot selection and forward–backward passes.
- Natural Gradient Boosting (NGBoost) employs a natural gradient to build a probabilistic estimation with remarkably greater accuracy.

- Random Forest (RF) uses a bagging approach to create an ensemble of DT, resulting in a reliable predictive model.
- Recurrent Neural Networks (RNN) use recurrent connections, hidden states, activation functions, and training through back-propagation.
- Support Vector Regression (SVR) can be used to accomplish regression analysis. Its primary idea is to find the best hyperplane that fits most points, minimizing the error.
- Wavelet Neural Networks (WNN) use the mother wavelet to captivate information from primary data and disintegrate it further.

1.2 Representative Applications of AI

The role of AI in Engineering, Science, Management, and other domains is rapidly expanding over time due to its capability to handle complex relationships between different features, uncertainty in data, and fewer data requirements (than in the traditional models). Representative applications include.

Customer Choices (Hu et al., 2023; Salminen et al., 2023)

Energy (Szczepaniuk & Szczepaniuk, 2023)

*Financial framework (*Bahoo et al., 2024*)*

Inventory (Chopra & Sharma, 2021; Li et al., 2023)

Knowledge-based management (Jarrahi et al., 2023; Taherdoost & Madanchian, 2023)

Manufacturing (Mypati et al., 2023; Naz et al., 2023; Plathottam et al., 2023)

Health (Castiglioni et al., 2021; Pesapane et al., 2018; Wang et al., 2021)

Precision agriculture (Son et al., 2024)

Quality Engineering (Martín et al., 2023; Aldoseri et al., 2023)

Robotics and Automation (Sarker, 2022; Soori et al., 2023).

Recent AI applications include text mining and Natural Language Processing. More details are presented in Chap. 6.

1.3 Scopus Analysis of AI

Scopus is an abstract and citation database launched in 2004 by Elsevier. It utilizes data analytics to understand various aspects of research papers, including related metrics (Scopus Content, 2024), and can be accessed using the following simple process: type 'www.scopus.com' in Internet Explorer. The dialog box appears, with two critical blocks: Search within and Search documents. In the Search within option, the user can find many sub-options to explore.

In summary, the accessibility of computational resources eventually accelerated the growth of AI. However, the user is expected to have complete domain knowledge of the mathematical framework before applying it, which is the motto of the present book.

1.4 Organization

There are seven chapters in the book.

This chapter briefly introduces AI, ML, and DL and provides the book's workflow. Chapter 2 discusses performance indicators, which judge the simulating ability of models and related software. Chapter 3 describes classical ML models used for forecasting and classification purposes: ANN, WNN, SVR, ELM, LR, and KNN. Activation functions are also part of this chapter. Chapter 4 describes a few advanced ML algorithms specifically used for forecasting. These are CNN, RNN, LSTM, Bi-directional (Bi)-LSTM, Gated Recurrent Units (GRU), and possible hybridizations of these algorithms. In addition, boosting algorithms, AdaBoost, XGBoost, and CatBoost are part of this chapter. Chapter 5 provides insight into fuzzification, defuzzification, FIS, ANFIS, Fuzzy Cognitive Mapping (FCM), optimization, and its fuzzy extension. It also comprises fuzzy-based CNN and LSTM and their hybridization, i.e., fuzzy CNN-LSTM. Chapter 6 discusses Blockchain, Advanced ML Techniques, Advanced Optimization Techniques, AI Tools, the Internet of Things (IoT), Big Data, Decision Support Systems (DSS), Taguchi Design of Experiments, Data Augmentation, and Cross-Validation. Chapter 7 presents representative case studies in Civil, Chemical, Mechanical, Electronics and Computer Science, Engineering, and Management. The purpose is to facilitate a comprehensive view regarding the applicability and potentiality of the methods. The appendix lists representative AI tools, data sources, books, and journals on AI. Figure 1.2 presents the flow of the topics in the chapters.

Before moving further, brief terminology on the theme of the book is presented below for the benefit of the readers:

Activation function facilitates non-linearity in the network, allowing it to capture complex patterns.

Architecture: It provides a holistic view of the network.

Batch size: The number of training examples used in one iteration.

Dropout: It is a regularization approach.

Epoch: It represents one cycle passing through the training dataset.

Error (Loss function): It is the deviation between simulated and observed datasets. The error can be positive or negative. It can also be expressed as squared deviation. Note that the purpose of any modelling approach is to minimize discrepancies.

Forecasting: It is a statistical approach to predict output.

Learning rate is the pace at which the updation of parameters can occur.

Momentum factor: It minimizes deviations in weights and enriches the training mechanism.

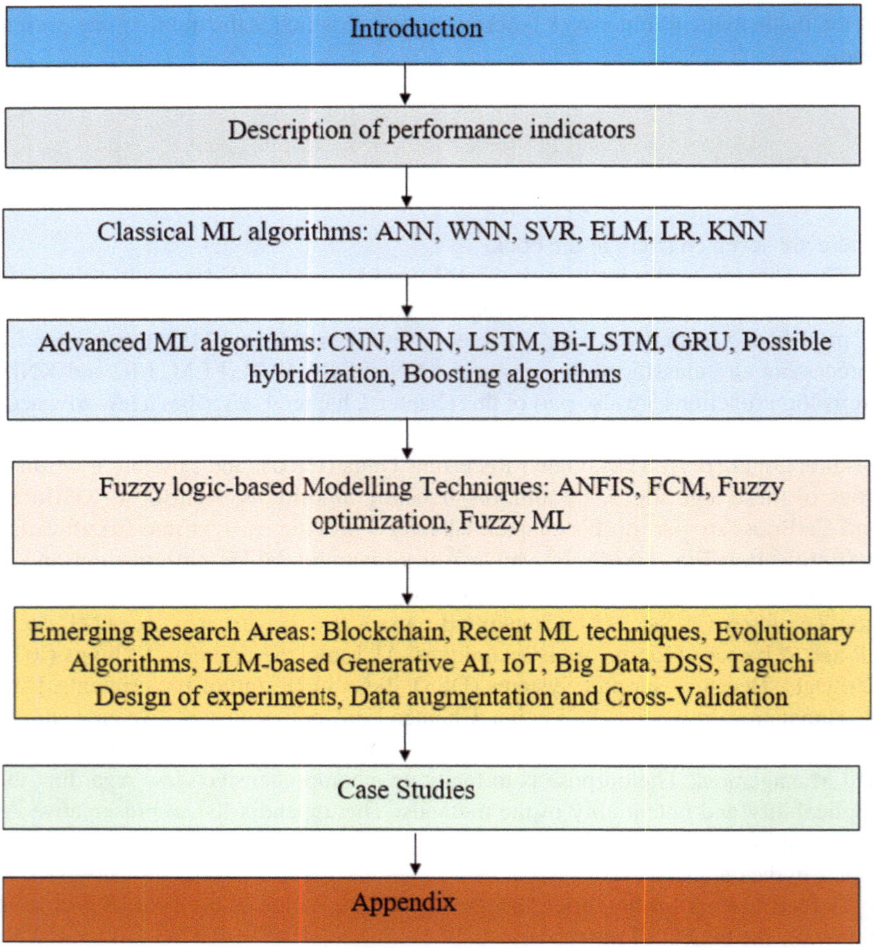

Fig. 1.2 The flow of the topics in the chapters

Parameters and hyperparameters: Internal elements of the model are termed as parameters. Hyperparameters are those parameters that significantly influence the training process.

Pre-processing: It is a preliminary step before transmitting the data to the network, such as normalization, outlier, and identification.

Supervised learning: It relates input and labelled output during training. In contrast, unsupervised learning does not have labelled output.

Training: A mechanism to determine optimum parameters (including connection strengths) that minimize the error between simulated and observed data.

Validation: Ideally, a model should not be tested on the dataset trained earlier, mainly for unbiased evaluation. For example, monthly rainfall and runoff data are available for 1970–2020. Out of which, data from 1970 to 2010 can be utilized for training. Accordingly, weights and other parameters are established. These and the last 10 years of rainfall are used to compute runoff for possible comparison with the observed data, as a part of testing.

Weights are connection strengths that vary from $(-\infty,\infty)$ and are continuously updated in the training process.

Exercise problems and advanced review questions are part of Chaps. 2 to 6. Algorithms, models, and techniques are used interchangeably in the book: data, points, and datasets; loss function, error, and discrepancy. It is requested to note the same.

Revision Questions

1.1 What is AI? What is the mechanism behind the same?
1.2 What are strong and weak AI?
1.3 What are the sub-categories of AI?
1.4 What is the philosophy of ML and DL?
1.5 What is the significant difference between ML and DL? Compare with two features.
1.6 Mention applications of AI in engineering and management.
1.7 What is Scopus analytics? How is it useful?
1.8 What are the challenges of applying AI to real-world problems?

Advanced Review Questions

1.9 Mention three views of researchers about AI.
1.10 Can you only employ DL for every chosen problem instead of ML? If yes or no, justify your view.
1.11 Can you make inferences from a few case studies where ML and DL were applied?
1.12 Do you think AI is necessary in engineering and management? Justify!

References

Alaskar, H., & Saba, T. (2021). Machine learning and deep learning: A comparative review. In K.K. Singh Mer, V. B. Semwal, V. Bijalwan, & R. G. Crespo (Eds.), *Proceedings of Integrated Intelligence Enable Networks and Computing*. Algorithms for Intelligent Systems. Springer, Singapore, 143–150.

Aldoseri, A., Al-Khalifa, K. N., & Hamouda, A. M. (2023). Re-thinking data strategy and integration for artificial intelligence: concepts, opportunities, and challenges. *Applied Sciences, 13*, 7082.

Bahoo, S., Cucculelli, M., Goga, X., & Mondolo, J. (2024). Artificial intelligence in finance: A comprehensive review through bibliometric and content analysis. *SN Business & Economics, 4*, 23.

Castiglioni, I., Rundo, L., Codari, M., Leo, G. D., Salvatore, C., Interlenghi, M., Gallivanone, F., Cozzi, A., D'Amico, N. C., & Sardanelli, F. (2021). AI applications to medical images: from machine learning to deep learning. *Physica Medica, 83*, 9–24.

Chopra, R., & Sharma, G. D. (2021). Application of artificial intelligence in stock market forecasting: A critique, review, and research agenda. *Journal of Risk and Financial Management, 14*, 526.

Hu, X., Liu, A., Li, X., Dai, Y., & Nakao, M. (2023). Explainable AI for customer segmentation in product development. *CIRP Annals, 72*, 89–92.

Janiesch, C., Zschech, P., & Heinrich, K. (2021). Machine learning and deep learning. *Electron Markets, 31*, 685–695.

Jarrahi, M. H., Askay, D., Eshraghi, A., & Smith, P. (2023). Artificial intelligence and knowledge management: A partnership between human and AI. *Business Horizons, 66*, 87–99.

Li, X., Sigov, A., Ratkin, L., Ivanov, L. A., & Li, L. (2023). Artificial intelligence applications in finance: A survey. *Journal of Management Analytics, 10*, 676–692.

Martín, L., Sánchez, L., Lanza, J., & Sotres, P. (2023). Development and evaluation of artificial intelligence techniques for iot data quality assessment and curation. *Internet of Things, 22*, 100779.

Mypati, O., Mukherjee, A., Mishra, D., Pal, S. K., Chakrabarti, P. P., & Pal, A. (2023). A critical review on applications of artificial intelligence in manufacturing. *Artificial Intelligence Review, 56*(Suppl 1), 661–768.

Naz, F., Kumar, A., Agrawal, R., Garza-Reyes, J. A., Majumdar, A., & Chokshi, H. (2023). Artificial intelligence as an enabler of quick and effective production repurposing: An exploratory review and future research propositions. *Production Planning & Control*, 1–24.

Nilsson, N. J. (2009). *The quest for artificial intelligence: A history of ideas and achievements.* Cambridge University Press.

Pesapane, F., Codari, M., & Sardanelli, F. (2018). Artificial intelligence in medical imaging: Threat or opportunity? Radiologists again at the forefront of innovation in medicine. *European Radiology Experimental, 2*, 35.

Plathottam, S. J., Rzonca, A., Lakhnori, R., & Iloeje, C. O. (2023). A Review of artificial intelligence applications in manufacturing operations. *Journal of Advanced Manufacturing and Processing, 5*, e10159.

Russell, S., & Norvig, P. (2010). Artificial intelligence: A modern approach, Pearson.

Salminen, J., Mustak, M., Sufyan, M., & Jansen, B. J. (2023). How can algorithms help in segmenting users and customers? *A Systematic Review and Research Agenda for Algorithmic Customer Segmentation, Journal of Marketing Analytics, 11*, 677–692.

Sarker, I. H. (2022). AI-based modeling: techniques, applications and research issues towards automation, intelligent and smart systems. *SN Computer Science, 3*, 158.

Scopus Content (2024). https://www.elsevier.com/en-in/products/scopus/content. Accessed September 30, 2024

Son, N., Chen, C. R., & Syu, C. H. (2024). Towards artificial intelligence applications in precision and sustainable agriculture. *Agronomy, 14*, 239.

Soori, M., Arezoo, B., & Dastres, R. (2023). Artificial intelligence, machine learning and deep learning in advanced robotics, A review. *Cognitive Robotics, 3*, 54–70.

Szczepaniuk, H., & Szczepaniuk, E. K. (2023). Applications of artificial intelligence algorithms in the energy sector. *Energies, 16*, 347.

Taherdoost, H., & Madanchian, M. (2023). Artificial intelligence and knowledge management: Impacts, benefits, and implementation. *Computers, 12*, 72.

Wang, S., Cao, G., Wang, Y., Liao, S., Wang, Q., Shi, J., Li, C., & Shen, D. (2021). Review and prospect: Artificial intelligence in advanced medical imaging. *Frontiers in Radiology, 1*, 781868.

Suggested Further Reading

Chang, F. J., Chang, L. C., & Chen, J. F. (2023). Artificial intelligence techniques in hydrology and water resources management. *Water, 15*, 1846.

Chapter 2
Description of Performance Indicators

2.1 Introduction

The present chapter briefly discusses representative performance indicators used to examine the competence of the chosen ML algorithm in simulating observed data (Jackson et al., 2019). Binary classification-based indicators are also part of this chapter.

2.2 Performance Indicators

Mathematical descriptions of indicators are as follows.

Let x_i, y_i represent observed and simulated values, respectively. μ_x, μ_y are mean of x_i, y_i *and* σ_x, σ_y *are* corresponding standard deviations. N is the number of datasets.

a. The Sum of the Square Loss Function (SSLF) is (Eq. 2.1):

$$\text{SSLF} = \sum_{i=1}^{\text{N}} (x_i - y_i)^2 \tag{2.1}$$

b. Mean Square Loss Function (MSLF) is the mean of SSA (Eq. 2.2):

$$\text{MSLF} = \frac{1}{\text{N}} \times \text{SSLF} \tag{2.2}$$

c. Root Mean Square Loss Function (RMSLF) is (Eq. 2.3):

$$\text{RMSLF} = \sqrt{\text{MSLF}} \tag{2.3}$$

K. Srinivasa Raju and D. Nagesh Kumar, *Artificial Intelligence and Machine Learning Techniques in Engineering and Management*,
https://doi.org/10.1007/978-981-96-2621-2_2

d. Normalized Root Mean Square Loss Function (NRMSLF) is (Eq. 2.4):

$$\mathrm{NRMSLF} = \frac{\mathrm{RMSLF}}{\mu_x} \tag{2.4}$$

e. Average Absolute Relative Loss Function (AARLF) is the ratio of the absolute Loss Function to the observed value (Eq. 2.5):

$$\mathrm{AARLF} = \frac{1}{\mathrm{N}}\sum_{i=1}^{\mathrm{N}}\left|\frac{(y_i - x_i)}{x_i}\right| \tag{2.5}$$

f. Normalized Standard Loss Function (NSLF) is (Eq. 2.6):

$$\mathrm{NSLF} = \frac{\sigma_y}{\sigma_x} \tag{2.6}$$

g. The coefficient of Correlation R (or CC) is a regression measure mainly in a linear mapping framework between simulated and observed (Eq. 2.7). It provides how the regression line best characterizes the data.

$$\mathrm{R} = \frac{\sum_{i=1}^{\mathrm{N}}(x_i - \mu_x)(y_i - \mu_y)}{(\mathrm{N}-1)\sigma_x\sigma_y} \tag{2.7}$$

h. Nash Sutcliffe Efficiency (NSE) is (Nash & Sutcliffe, 1970) (Eq. 2.8):

$$\mathrm{NSE} = 1 - \frac{\sum_{i=1}^{\mathrm{N}}(x_i - y_i)^2}{\sum_{i=1}^{\mathrm{N}}(x_i - \mu_x)^2} \tag{2.8}$$

i. Kling Gupta Efficiency (KGE) is based on correlation, variability, and mean biases (Gupta et al., 2009) (Eq. 2.9):

$$\mathrm{KGE} = 1 - \left[\left((\mathrm{R}-1)^2 + \left(\frac{\sigma_y}{\sigma_x} - 1\right)^2 + \left(\frac{\mu_y}{\mu_x} - 1\right)^2\right)\right]^{0.5} \tag{2.9}$$

j. Taylor Skill Score (TSS) is based on a standard deviation of observed and simulated spatial correlation coefficient R (Taylor, 2001) (Eq. 2.10):

$$\mathrm{TSS} = \frac{4(1+\mathrm{R})^4}{\left(1+\mathrm{R}_0^4\right)\left(\frac{\sigma_y}{\sigma_x} + \frac{\sigma_x}{\sigma_y}\right)^2} \tag{2.10}$$

R_0 is the highest possible R based on the perception of the user.

k. Fractional Skill Score (FSS) (Ma et al., 2018; Roberts & Lean, 2008) is (Eq. 2.11):

$$\mathrm{FSS} = 1 - \frac{\left(\frac{1}{\mathrm{N}}\sum_{i=1}^{\mathrm{N}}(x_i - y_i)^2\right)}{\left(\frac{1}{\mathrm{N}}\sum_{i=1}^{\mathrm{N}}x_i^2\right) + \left(\frac{1}{\mathrm{N}}\sum_{i=1}^{\mathrm{N}}y_i^2\right)} \quad (2.11)$$

Lower values (ideally zero) are preferred for indicators (a–f). Higher values (ideally one) are preferred for indicators (g–k). Detailed information about some of the indicators and their ranges is available from Moriasi et al. (2007) and Moriasi et al. (2015).

Numerical problem 2.1. Table 2.1 presents the observed and simulated rainfall obtained by the ML algorithm. Compute indicators described in this section.

Solution:

Related calculations are showcased in Table 2.1.

$$\text{Number of datasets} = 10.$$

$$\sum x = 121\ \text{cm}, \sum y = 115\ \text{cm}, \sum xy = 1450\ \text{cm}^2$$

$$\sum x^2 = 1525\ \text{cm}^2, \sum y^2 = 1487\ \text{cm}^2$$

$$\mu_x = 12.1\ \text{cm}, \mu_y = 11.5\ \text{cm}.$$

$$\sigma_x = 2.6013\ \text{cm}, \sigma_y = 4.2753\ \text{cm}.$$

Table 2.2 presents a list of indicators and corresponding values based on Eqs. 2.1–2.11.

Numerical problem 2.2. Table 2.3 presents the number of times noise beyond a certain decibels (dB) was measured using electromagnetic sensors in the manufacturing industry. A simulated number of occurrences from modelling is also part of Table 2.3. Compute NSE, CC, KGE, TSS, and FSS. Present this information using a bar chart.

Solution:

Related calculations are presented in Table 2.3.

Table 2.4 and Fig. 2.1 present indicators and corresponding values.

Table 2.1 Observed and simulated values and related calculations

Dataset	Observed rainfall x (cm)	Simulated rainfall y (cm)	$(x-\mu_x)^2$ (cm^2)	$(y-\mu_y)^2$ (cm^2)	$(x-y)^2$ (cm^2)	$\left\lvert\frac{y-x}{x}\right\rvert$ (No unit)	$(x-\mu_x)(y-\mu_y)$ (cm^2)	xy (cm^2)	x^2 (cm^2)	y^2 (cm^2)
1	14	16	3.61	20.25	4	0.1429	8.55	224	196	256
2	12	8	0.01	12.25	16	0.3333	0.35	96	144	64
3	9	8	9.61	12.25	1	0.1111	10.85	72	81	64
4	14	8	3.61	12.25	36	0.4286	−6.65	112	196	64
5	16	13	15.21	2.25	9	0.1875	5.85	208	256	169
6	11	14	1.21	6.25	9	0.2727	−2.75	154	121	196
7	12	10	0.01	2.25	4	0.1667	0.15	120	144	100
8	13	15	0.81	12.25	4	0.1538	3.15	195	169	225
9	13	18	0.81	42.25	25	0.3846	5.85	234	169	324
10	7	5	26.01	42.25	4	0.2857	33.15	35	49	25

Table 2.2 Indicators and corresponding values

Indicator	Value	Unit
SSLF	112	cm^2
MSLF	11.2	cm^2
RMSLF	3.3466	cm
NRMSLF	0.2766	No unit
AARLF	0.2467	No unit
NSLF	1.6435	No unit
R (or CC)	0.5845	No unit
NSE	−0.8391	No unit
KGE	0.2324	No unit
TSS	0.3107	No unit
FSS	0.9628	No unit

2.3 Indicators in Binary Classification Problems

Many indicators also exist in binary classification problems. Before moving into the detailed understanding of indicators, the following related definitions will be helpful.

Confusion matrix: It is a matrix with four different possibilities of simulated and observed data (Fig. 2.2), and the related description is as follows:

- True positive (TP): Envisaged as positive, and it is correct. Example: It is envisaged that a flood will occur, and the flood has occurred.
- False positive (FP): Envisaged as positive, and it is incorrect. Example: It is envisaged that a flood will occur, and the flood has not occurred (falling under Type 1 error).
- False Negative (FN): Envisaged as negative, and it is incorrect. Example: It is envisaged that a flood will not occur, and the flood has occurred (falling under Type 2 error).
- True Negative (TN): Envisaged as negative, and it is correct. Example: It is envisaged that a flood will not occur, and the flood has not occurred

Some of the standard classification indicators derived from the confusion matrix are as follows (Agrawal, 2023; Czakon, 2023) (Eqs. 2.12–2.17):

$$\text{True Positive and False Positive Rates (TPR, FPR)} = \frac{\text{TP}}{\text{TP} + \text{FN}}, \frac{\text{FP}}{\text{TN} + \text{FP}} \tag{2.12}$$

$$\text{True Negative and False Negative Rates (TNR, FNR)} = \frac{\text{TN}}{\text{TP} + \text{FP}}, \frac{\text{FN}}{\text{TP} + \text{FN}} \tag{2.13}$$

$$\text{Precision } (P) = \frac{\text{TP}}{\text{TP} + \text{FP}} \tag{2.14}$$

Table 2.3 Observed and simulated values and related calculations

Dataset	Number of times noise (beyond a certain decibels) observed by sensor x	Simulated number of times by modelling y	$(x-\mu_x)^2$	$(y-\mu_y)^2$	$(x-y)^2$	$\left\|\frac{y-x}{x}\right\|$	$(x-\mu_x)(y-\mu_y)$	xy	x^2	y^2
1	32	20	0.09	16	144	0.3750	1.2	640	1024	400
2	42	28	94.09	16	196	0.3333	38.8	1176	1764	784
3	21	20	127.69	16	1	0.0476	45.2	420	441	400
4	72	56	1576.09	1024	256	0.2222	1270.4	4032	5184	3136
5	28	22	18.49	4	36	0.2143	8.6	616	784	484
6	24	16	68.89	64	64	0.3333	66.4	384	576	256
7	36	30	13.69	36	36	0.1667	22.2	1080	1296	900
8	28	14	18.49	100	196	0.5000	43	392	784	196
9	22	12	106.09	144	100	0.4545	123.6	264	484	144
10	18	22	204.49	4	16	0.2222	28.6	396	324	484

Table 2.4 Indicators and corresponding values

Indicator	Value
NSE	0.5310
CC	0.9252
KGE	0.6656
TSS	0.8170
FSS	0.9473

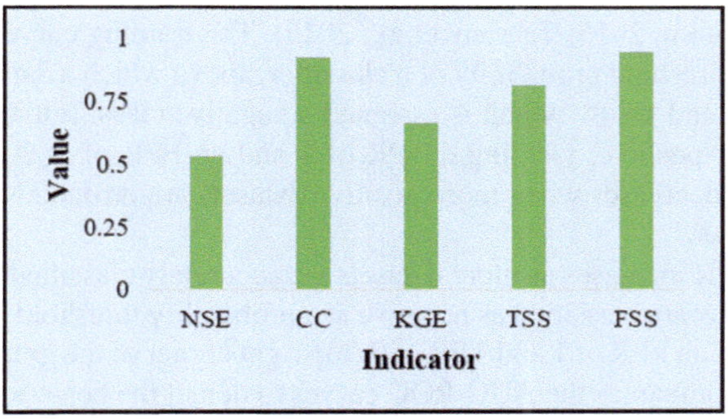

Fig. 2.1 Selected indicators and their values

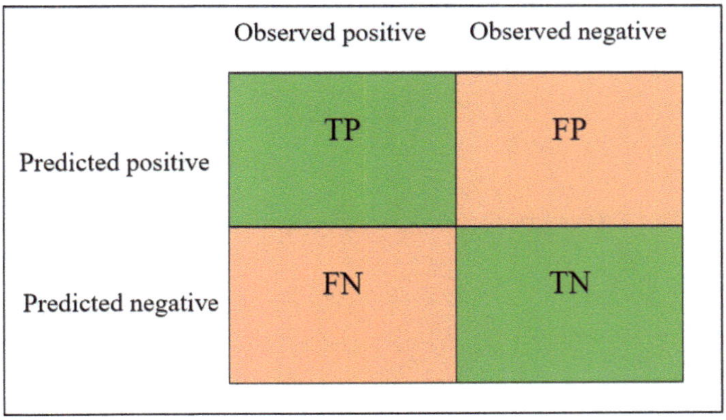

Fig. 2.2 Confusion matrix

$$\text{F-measure} = \frac{2 \times \text{TPR} \times \text{P}}{\text{TPR} + \text{P}} \tag{2.15}$$

$$\text{Accuracy} = \frac{\text{TP} + \text{TN}}{\text{TP} + \text{TN} + \text{FP} + \text{FN}} \tag{2.16}$$

$$\text{Error} = 1 - \text{Accuracy} \tag{2.17}$$

The value of TPR gives an idea of how many correct positive results are present in all of the positive scenarios. On the other hand, FPR shows how many incorrect positive examples are present in all negative scenarios.

Area Under the Curve-Receiver Operating Characteristic (AUC-ROC) curve conveys the change in the classification ability of the model with various thresholds. Here, FPR and TPR are shown on the x and y axes (Fig. 2.3) (Chapi et al., 2017; Madhuri et al., 2021).

Users can plot an AUC-ROC curve on a graph sheet (pairs of FPR and TPR for various thresholds). It can be used to compare and validate various algorithms (Shahabi & Hashim, 2015; Tehrany et al., 2015). The training curve is constructed by varying the threshold probability of a classifier, above which a dataset is assigned a positive class and below which is assigned a negative class. Initially, all datasets are classified as positive, yielding a TPR of 1 and an FPR of 0. As the threshold increases, TPR decreases when more positive datasets are erroneously classified as the negative class.

Similarly, FPR increases as more datasets are accurately classified as the negative class. All datasets are classified as negative at a probability threshold of 1, giving the other extreme of an FPR of 1 and TPR of 0 (top right corner of the graph). A diagonal line usually accompanies the AUC-ROC curves to depict the behaviour of the worst possible classifier (i.e., a random classifier that correctly predicts 50% of the time)

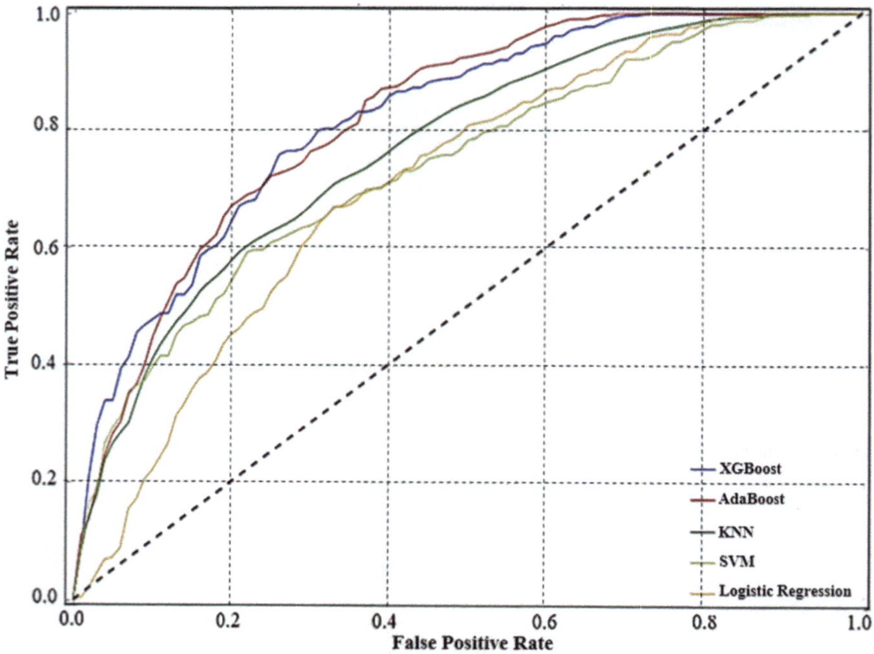

Fig. 2.3 Representative AUC-ROC training curves for different algorithms (Modified and adapted from Madhuri et al., (2021) under CC BY-NC-ND 4.0 license)

(refer to Fig. 2.3). The greater distance of the AUC-ROC curve from the diagonal line indicates the model's efficacy in discriminating the classes, which occurs in the high AUC-ROC situation, resulting in the arch curve. The range of AUC-ROC is between 0 to 1. Generally, an AUC-ROC of 1 ideally distinguishes between negative and positive classes. If the value is 0, the model predicts positive classes as negative and vice-versa (Madhuri, 2022).

As a note, it is the choice of the individual to pick up the relevant indicator according to their requirements.

Numerical problem 2.3. Two situations exist in water distribution networks (WDN): leak and no-leak. During observation by the field engineers, 12 leaks and 16 no-leaks were observed. Later, they used one of the algorithms to simulate these types of leaks. The algorithm identified only 9 leaks (TP), 12 no-leaks (TN), 4 no-leaks as leaks (FP), and 3 leaks as no-leaks (FN). Plot a confusion matrix for the given data. Analyze the problem with standard classification indicators.

Solution:

Confusion matrix for $\mathrm{TP} = 9,\ \mathrm{TN} = 12, \mathrm{FP} = 4,\ \mathrm{FN} = 3$ is presented as Fig. 2.4

$$\mathrm{TPR} = \frac{\mathrm{TP}}{\mathrm{TP} + \mathrm{FN}} = \frac{9}{9+3} = 0.75$$

$$\mathrm{FPR} = \frac{\mathrm{FP}}{\mathrm{TN} + \mathrm{FP}} = \frac{4}{12+4} = 0.25$$

$$\mathrm{TNR} = \frac{\mathrm{TN}}{\mathrm{TP} + \mathrm{FP}} = \frac{12}{9+4} = 0.923$$

$$\mathrm{FNR} = \frac{\mathrm{FN}}{\mathrm{TP} + \mathrm{FN}} = \frac{3}{9+3} = 0.25$$

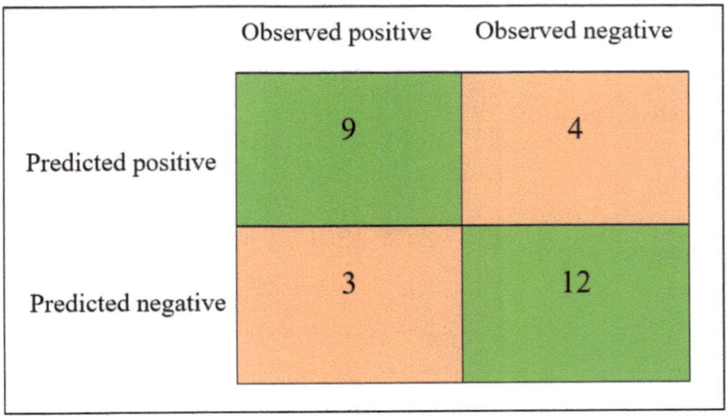

Fig. 2.4 Confusion matrix for the given numerical problem

$$\text{Precision} = \frac{\text{TP}}{\text{TP} + \text{FP}} = \frac{9}{9+4} = 0.69231$$

$$\text{F-measure} = \frac{2 \times \text{TPR} \times \text{P}}{\text{TPR} + \text{P}} = \frac{2 \times 0.75 \times 0.69231}{0.75 + 0.69231} = 0.72$$

$$\text{Accuracy} = \frac{\text{TP} + \text{TN}}{\text{TP} + \text{TN} + \text{FP} + \text{FN}} = \frac{9+12}{9+12+4+3} = 0.75$$

$$\text{Error} = 1 - 0.75 = 0.25$$

Numerical problem 2.4. Two scenarios existed in students' performance in academic institutions: pass or fail. During analysis by the student welfare division, it was noted that 22 students passed and 20 failed. The official who is monitoring the process used a simulation algorithm. It identified 14 (TP), 14 (TN), 6 (FP), and 8 (FN). Draw a confusion matrix for the given data. Compute Precision, Specificity, F-measure, and Accuracy.

Solution:

Confusion matrix for TP = 14, TN = 14, FP = 6, FN = 8 is presented as Fig. 2.5

$$\text{Precision} = \frac{\text{TP}}{\text{TP} + \text{FP}} = \frac{14}{14+6} = 0.7$$

$$\text{FPR} = \frac{\text{FP}}{\text{TN} + \text{FP}} = \frac{6}{14+6} = 0.3$$

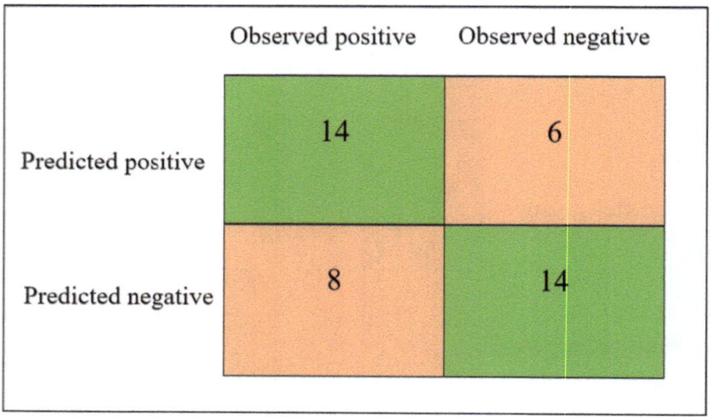

Fig. 2.5 Confusion matrix for the given numerical problem

Table 2.5 Data of FPR and TPR for various thresholds

Dataset	FPR	TPR
1	0	0
2	0	0.3
3	0	0.5
4	0	0.7
5	0.3	0.7
6	0.5	0.7
7	0.5	0.9
8	0.7	0.9
9	0.7	1
10	0.9	1
11	1	1

$$\text{TPR} = \frac{\text{TP}}{\text{TP} + \text{FN}} = \frac{14}{14 + 8} = 0.63636$$

$$\text{F-measure} = \frac{2 \times \text{TPR} \times \text{P}}{\text{TPR} + \text{P}} = \frac{2 \text{ x } 0.63636 \text{ x } 0.7}{0.63636 + 0.7} = 0.6667$$

$$\text{Accuracy} = \frac{\text{TP} + \text{TN}}{\text{TP} + \text{TN} + \text{FP} + \text{FN}} = \frac{14 + 14}{14 + 14 + 6 + 8} = 0.6667$$

Numerical problem 2.5. Table 2.5 consists of FPR and TPR obtained for various thresholds to predict the thermophysical properties of hybrid nanofluids. Draw the AUC-ROC curve and compute the indicator.

Solution: Refer to Fig. 2.6

AUC-ROC value $= 0.7 \times 0.5 + 0.9 \times 0.2 + 1 \times 0.3 = 0.83$ [accumulation of individual areas yields AUC-ROC value].

Representative Software for the Computation of Indicators

Agricultural and Meteorological software (https://agrimetsoft.com/calculators/) facilitates the calculation of various indicators.

Revision Questions and Exercise Problems

2.1 Why should simulated and observed be compared?
2.2 What is the role of indicators in modelling? Mention six indicators relevant to engineering with mathematical expressions and units.
2.3 What is TP, TN, FP, and FN? What is their purpose in modelling? What are the minimum and maximum values that are possible for these indicators?
2.4 What are Type 1 and Type 2 errors?
2.5 Discuss salient features of the AUC-ROC curve.

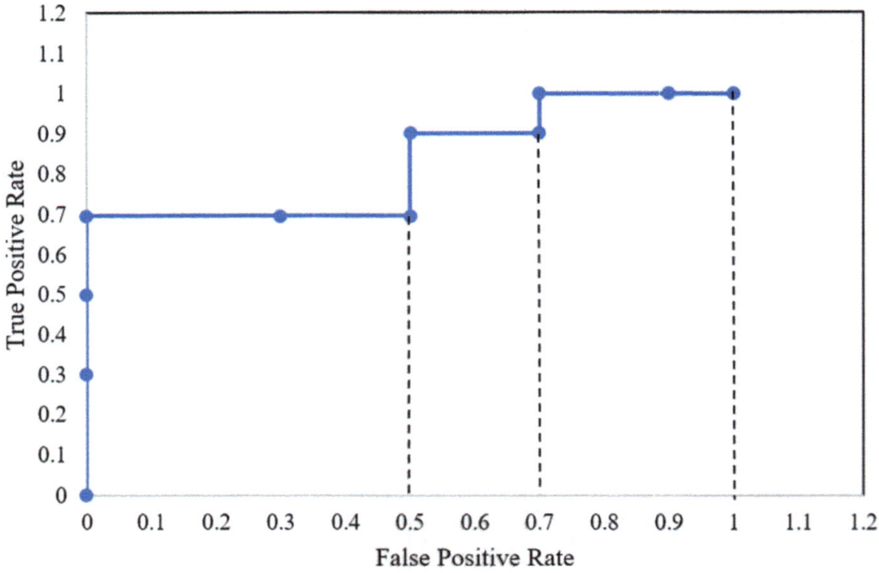

Fig. 2.6 AUC-ROC curve

2.6 What is the ideal value of the AUC-ROC curve?

2.7 Can you differentiate between accuracy and the AUC-ROC curve? Which do you prefer and why?

2.8 Information on the observed and simulated number of vehicles travelling on a highway for days 1–10 for a specific duration is as follows.
Observed number of vehicles: 130, 110, 80, 130, 150, 120, 110, 120, 120, 60
Simulated number of vehicles: 150, 80, 90, 100, 120, 130, 90, 140, 170, 40
Compute KGE, NSE, FSS and TSS. Draw the inferences from the obtained values.

2.9 The problem is related to Computer Numerical Control (CNC) machines, where the number of metallic sheets handled for four consecutive hours is 20, 35, 40, and 65. However, the machine is expected to handle 18, 40, 35, and 70 metallic sheets as per norms. Analyze Loss Function-based indicators.

2.10 The problem is related to the electronics engineering domain, where several smoke detection sensors were developed. Compute Precision and Accuracy for TP = 10, TN = 320, FP = 20, and FN = 50. Make relevant inferences.

2.11 Table 2.6 contains FPR and TPR obtained for various thresholds in sorption-enhanced biomass chemical looping gasification. Draw the AUC-ROC curve and compute the related value.

Table 2.6 Data of FPR and TPR for various thresholds

Dataset	FPR	TPR
1	0	0
2	0	0.35
3	0	0.45
4	0	0.55
5	0.35	0.75
6	0.55	0.80
7	0.6	0.85
8	0.7	0.9
9	0.8	1
10	0.85	1
11	1	1

Advanced Review Questions

2.12 Can you develop indicators other than those mentioned in this chapter? If so, what will be the advantages of the proposed indicators over the existing ones?
2.13 Do you think weights are to be assigned for the indicators? Justify your answer!
2.14 Do you have any challenges while computing the AUC-ROC value? If yes, what are they?

References

Agrawal SK (2023) Metrics to evaluate your classification model to take the right decisions. Accessed February 10, 2024, https://www.analyticsvidhya.com/blog/2021/07/metrics-to-evaluate-your-classification-model-to-take-the-right-decisions/

Chapi, K., Singh, V. P., Shirzadi, A., Shahabi, H., Bui, D. T., Pham, B. T., & Khosravi, K. (2017). A novel hybrid artificial intelligence approach for flood susceptibility assessment. *Environmental Modelling and Software, 95*, 229–245.

Czakon, J. (2023). 24 Evaluation metrics for binary classification (and when to use them). Accessed February 10, 2024, https://neptune.ai/blog/evaluation-metrics-binary-classification

Gupta, H. V., Kling, H., Yilmaz, K. K., & Martinez, G. F. (2009). Decomposition of the mean squared error and NSE performance criteria: Implications for improving hydrological modelling. *Journal of Hydrology, 377*, 80–91.

Jackson, E. K., Roberts, W., Nelsen, B., Williams, G. P., Nelson, E. J., & Ames, D. P. (2019). Introductory overview: Error metrics for hydrologic modelling—A review of common practices and an open source library to facilitate use and adoption. *Environmental Modelling & Software, 119*, 32–48.

Ma, S., Chen, C., He, H., Wu, D., & Zhang, C. (2018). Assessing the skill of convection-allowing ensemble forecasts of precipitation by optimization of spatial-temporal neighborhoods. *Atmosphere, 9*, 43.

Madhuri, R. (2022). *Risk assessment and mitigation strategies for urban floods under climate change*. Ph.D. thesis, BITS Pilani, India.

Madhuri, R., Sistla, S., & Raju, K.S. (2021). Application of machine learning algorithms for flood susceptibility assessment and risk management. *Journal of Water and Climate Change, 12*, 2608–2623 [open access paper under under CC BY-NC-ND 4.0 license].

Moriasi, D. N., Arnold, J. G., Liew, M. W. V., Bingner, R. L., Harmel, R. D., & Veith, T. L. (2007). Model evaluation guidelines for systematic quantification of accuracy in watershed simulations. *Transactions of the American Society of Agricultural and Biological Engineers, 50*, 885–900.

Moriasi, D. N., Gitau, M. W., Pai, N., & Daggupati, P. (2015). Hydrologic and water quality models: performance measures and evaluation criteria. *American Society of Agricultural and Biological Engineers, 58*, 1763–1785.

Nash, J. E., & Sutcliffe, J. V. (1970). River flow forecasting through conceptual models part I—A discussion of principles. *Journal of Hydrology, 10*, 282–290.

Roberts, N. M., & Lean, H. W. (2008). Scale-selective verification of rainfall accumulations from high-resolution forecasts of convective events. *Monthly Weather Review, 136*, 78–97.

Shahabi, H., & Hashim, M. (2015). Landslide susceptibility mapping using GIS-based statistical models and remote sensing data in tropical environment. *Scientific Reports, 5*, 9899.

Taylor, K. E. (2001). Summarizing multiple aspects of model performance in a single diagram. *Journal of Geophysical Research: Atmospheres, 106*(D7), 7183–7192.

Tehrany, M. S., Pradhan, B., Mansor, S., & Ahmad, N. (2015). Flood susceptibility assessment using GIS-based support vector machine model with different Kernel types. *CATENA, 125*, 91–101.

Suggested Further Reading

Hridik, P., Vasan, A., & Raju, K. S. (2022). Leak detection in water distribution networks using deep learning. *ISH Journal of Hydraulic Engineering, 29*, 674–682.

Chapter 3
Classical Machine Learning Algorithms

3.1 Introduction

The present chapter is a blend of classically familiar algorithms, namely, Artificial Neural Networks (ANN), Wavelet Neural Networks (WNN), Support Vector Regression (SVR), Extreme Learning Machine (ELM), Logistic Regression (LR), and K-Nearest Neighbour (KNN). Before proceeding to the details of these algorithms, an important topic, the activation function, is briefly discussed.

Note: Given input and output values are considered to be normalized in all the numerical problems discussed here and in other chapters. In addition, a representative situation for connecting the problem to the real-world scenario was provided.

3.2 Activation Function

The activation function estimate output from the given input (Fig. 3.1). These can be developed or modified depending on the requirement. Figure 3.2 presents selected activation functions and their mathematical philosophy (Sharma, 2017).

3.3 Artificial Neural Networks

ANN can develop a non-linear relation between inputs and outputs. Figure 3.3 presents architecture. It comprises layers. Each layer receives output from the preceding layers, i.e., output from each layer contributes to the next layer. Here, Feed-Forward with Back-Propagation (FFBP)-based ANN is discussed briefly (Fig. 3.3).

K. Srinivasa Raju and D. Nagesh Kumar, *Artificial Intelligence and Machine Learning Techniques in Engineering and Management*,
https://doi.org/10.1007/978-981-96-2621-2_3

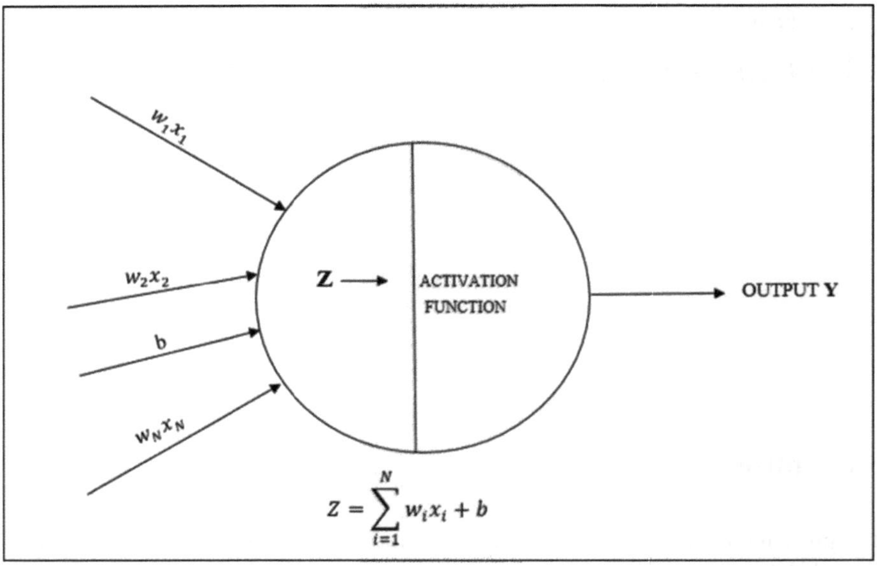

Fig. 3.1 Weighted sum input to the neuron (or node), activation function, and the resulting output

FFBP has forward and backward phases (Rao, 2000). The forward phase is associated with transmitting inputs to the output layer via the hidden layer using the activation function. Contrarily, the error between simulated and observed is propagated proportionately to the preceding layers in the backward phase. The process can be stopped when there is no change in the error between successive epochs (or termination criterion as specified by the user). Weights established at this stage are considered optimal and used for further analysis. The weight adjustment process is as follows (Eq. 3.1):

$$\Delta\omega_{ij}(n) = -\mathrm{L_r} \times \frac{\partial \mathrm{E}}{\partial \omega_{ij}} + \mathrm{M_r} \times \Delta\omega_{ij}(n-1) \tag{3.1}$$

where $\mathrm{M_r}$ and $\mathrm{L_r}$ are momentum and learning rates, respectively. $\Delta\omega_{ij}(n-1)$, $\Delta\omega_{ij}(n)$ are weight changes between nodes i and j in the course of $(n-1)$ and $n$epochs. The computation of updated weights is as follows (Eq. 3.2)

$$\omega_{ij}(\text{new}) = \omega_{ij}(\text{old}) + \Delta\omega_{ij} \tag{3.2}$$

Further simplifying Eqs. 3.1 and 3.2, without momentum factor yields (Eq. 3.3),

$$\omega_{jk}^{i}\text{new} = \omega_{jk}^{i}\text{old} + \mathrm{L_r} \times \mathrm{E}_k^{i+1} \times x_{jk} \tag{3.3}$$

<table>
<tr><th></th><th>Shape of activation function</th><th>Mathematical philosophy</th></tr>
<tr><td>(a)</td><td>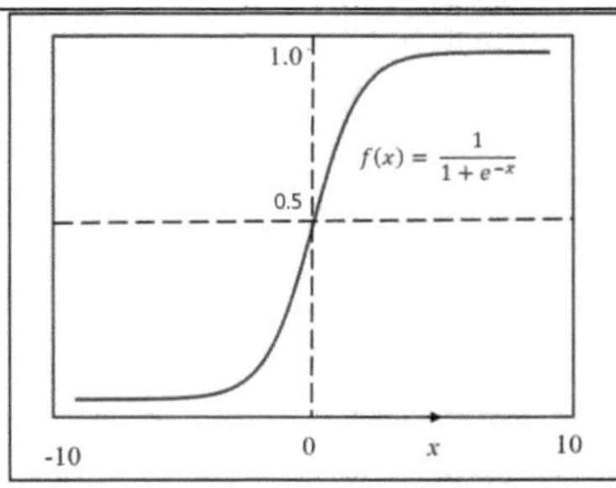
</td><td>The output is in the form of probability. S-shaped is continuously differentiable; the range of the function is 0 to 1.</td></tr>
<tr><td>(b)</td><td>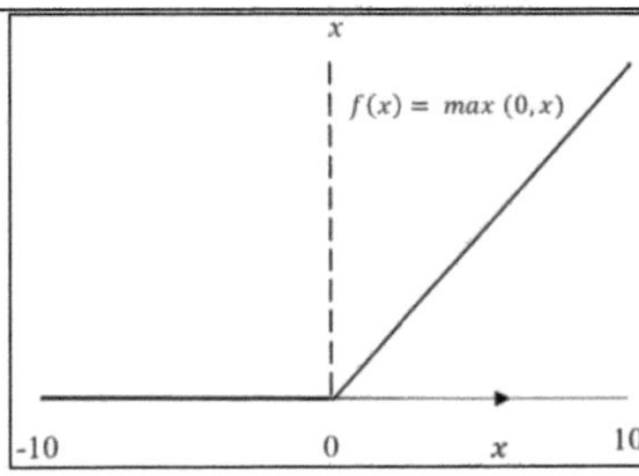
</td><td>A certain number of nodes only get activated, making it computationally more efficient; The range is 0 to ∞.</td></tr>
<tr><td>(c)</td><td>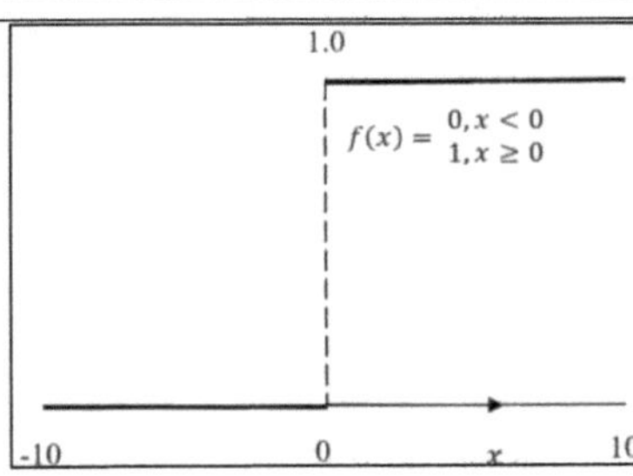
</td><td>Simple function, which works with the concept of the threshold. Not capable of handling multi-class clustering problems.</td></tr>
<tr><td>(d)</td><td>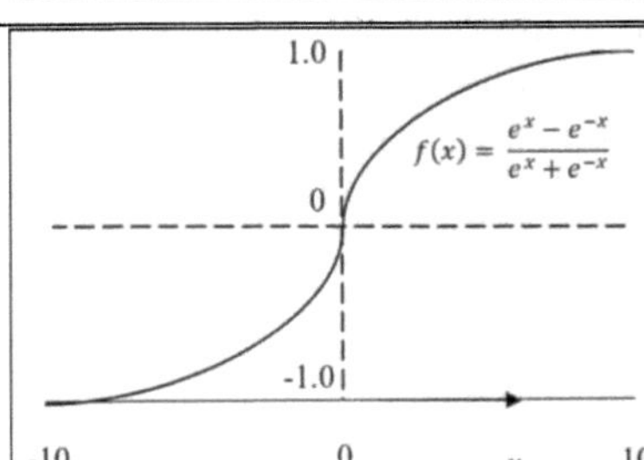
</td><td>Most flexible; the range is -1 to 1.</td></tr>
</table>

Fig. 3.2 Activation functions **a** Sigmoid **b** Rectified Linear Unit (ReLU) **c** Binary step function **d** and Hyperbolic Tangent and their mathematical philosophy

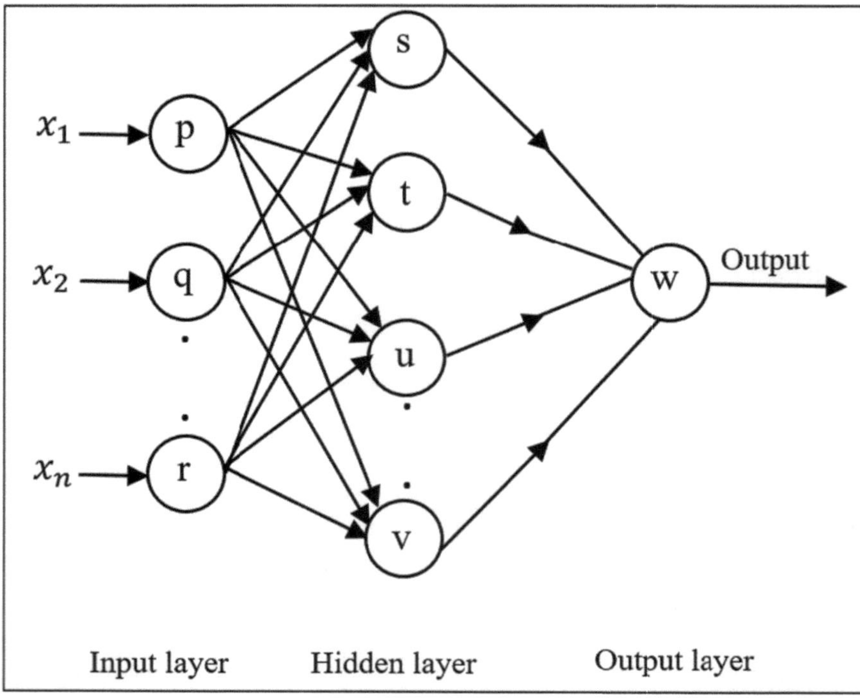

Fig. 3.3 The representative architecture of ANN

where ω^i_{jk} is weight relating ith layer, jth node to the $(i + 1)$th layer, kth node; E^{i+1}_k is the error of $(i + 1)$th layer, kth node; x_{jk} is the input from ith layer, jth node to the $(i + 1)$th layer, kth node.

Numerical problem 3.1. In a typical architecture, one input layer (with four inputs, namely, Dissolved solids, Electrical conductivity, Turbidity, and pH) and one hidden layer with one node exist. Input values are 0.2, 0.3, 0.4 and 0.5. Assume connection strengths between four inputs and the hidden layer as 0.4, 0.5, 0.6, and 0.7. Compute output from the hidden layer. Sigmoid activation function can be chosen. What may happen if the activation function is Hyperbolic Tangent?

Solution:

Weighted input to the hidden layer $\sum_{i=1}^{4} w_i x_i = 0.2 \times 0.4 + 0.3 \times 0.5 + 0.4 \times 0.6 + 0.5 \times 0.7 = 0.08 + 0.15 + 0.24 + 0.35 = 0.82$

Output from a hidden layer on the basis of Sigmoid function, $f(x) = \frac{1}{1+e^{-\sum wx}} = \frac{1}{1+e^{-0.82}} = 0.6942$

In the case of a Hyperbolic Tangent, output from the hidden layer, $f(x) = \frac{e^{\sum wx} - e^{-\sum wx}}{e^{\sum wx} + e^{-\sum wx}} = \frac{e^{0.82} - e^{-0.82}}{e^{0.82} + e^{-0.82}} = \frac{2.2704 - 0.4404}{2.2704 + 0.4404} = 0.675$

Output based on the Hyperbolic Tangent activation function is slightly lower compared to Sigmoid.

Numerical problem 3.2. Three inputs, current, sampling time, and temperature (A, B, C) with magnitudes 2, 4, and 6, produce an output X (state of health of battery) of 0.6. One hidden layer with two nodes, K and L, is suggested. Initial weights from input (A, B, C) to the K node are 0.1, 0.2, and 0.3; these values for the L node are 0.15, 0.25, and 0.35. Initial weights from hidden nodes K and L to output X are 0.4 and 0.45, respectively. The learning rate is 0.1. Use the Sigmoid activation function. Draw the architecture. Establish a relationship using FFBP-ANN and show the computations for one epoch.

Solution:

Figure 3.4 presents the architecture for the given problem.
Input values x_1, x_2, x_3 (Nodes A, B, C) are 2, 4, 6; Observed value O_X is 0.6
Activation function: Sigmoid $f(x) = \frac{1}{1+e^{-x}}$

Epoch 1:

Weighted input to the K node of the hidden layer $= \sum_{i=1}^{3} \omega_i x_i = 0.1 \times 2 + 0.2 \times 4 + 0.3 \times 6 = 2.8$
Output from the K node of the hidden layer, $O_K = \frac{1}{1+e^{-x}} = \frac{1}{1+e^{-2.8}} = 0.9427.$
Weighted input to the L node of the hidden layer$= \sum_{i=1}^{3} \omega_i x_i = 0.15 \times 2 + 0.25 \times 4 + 0.35 \times 6 = 3.4$
Output from the L node of the hidden layer, $O_L = \frac{1}{1+e^{-x}} = \frac{1}{1+e^{-3.4}} = 0.9677$
Weighted input to the output layer X, from K and L nodes $= 0.4 \times 0.9427 + 0.45 \times 0.9677 = 0.8125$

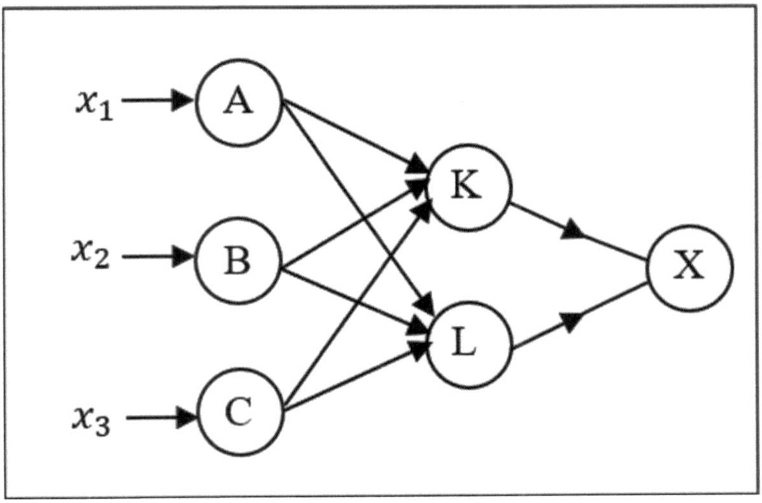

Fig. 3.4 The architecture of ANN for the given numerical problem

Predicted output P from the output layer X, $P_X = \frac{1}{1+e^{-x}} = \frac{1}{1+e^{-0.8125}} = 0.6927$
Observed output $O_X = 0.6$
Error $E_X = O_X - P_X = 0.6 - 0.6927 = -0.0927$
The division of errors is $E_{j-1} = O_n(1 - O_n) \sum \omega_{nj} E_j$ [If j is X, $j - 1$ is K, L]
Entrusting errors to elements in layer 2

$$\begin{aligned} E_K &= O_K \times (1 - O_K) \times (\omega_{KX} \times E_X) \\ &= 0.9427 \times (1-0.9427) \times (0.4 \times -0.0927) = -0.002003 \end{aligned}$$

$$\begin{aligned} E_L &= O_L \times (1 - O_L) \times (\omega_{LX} \times E_X) \\ &= 0.9677 \times (1-0.9677) \times (0.45 \times -0.0927) = -0.001304 \end{aligned}$$

Weights are updated as per Eq. 3.3. Tables 3.1 and 3.2 comprise updated weights connecting nodes in layers 1 and 2 & 2 and 3, respectively.

Numerical problem 3.3. The architecture is 5-4-1 (refer to Fig. 3.5). Five inputs (muscle strength, aerobic endurance, body mass index, speed, and flexibility) (A to E), with magnitudes 5, 6, 7, 8, and 9 affecting non-academic performance X, with the

Table 3.1 Updated weights connecting nodes in layers 1 and 2

Updated weight for input to hidden nodes	Equation for updating weights	Updated weights
AK	$\omega_{AK.old} + L_r \times E_K \times x_1$	$0.1 + 0.1 \times (-0.002003) \times 2 = 0.0996$
BK	$\omega_{BK.old} + L_r \times E_K \times x_2$	$0.2 + 0.1 \times (-0.002003) \times 4 = 0.1992$
CK	$\omega_{CK.old} + L_r \times E_K \times x_3$	$0.3 + 0.1 \times (-0.002003) \times 6 = 0.2988$
AL	$\omega_{AL.old} + L_r \times E_L \times x_1$	$0.15 + 0.1 \times (-0.001304) \times 2 = 0.1497$
BL	$\omega_{BL.old} + L_r \times E_L \times x_2$	$0.25 + 0.1 \times (-0.001304) \times 4 = 0.2495$
CL	$\omega_{CL.old} + L_r \times E_L \times x_3$	$0.35 + 0.1 \times (-0.001304) \times 6 = 0.3492$

Table 3.2 Updated weights connecting nodes in layers 2 and 3

Updated weight for the hidden to output nodes	Equation for updating weights	Updated weights
KX	$\omega_{KX.old} + L_r \times E_K \times E_X$	$0.4 + 0.1 \times (-0.002003) \times (-0.0927) = 0.4000188$
LX	$\omega_{LX.old} + L_r \times E_L \times E_X$	$0.45 + 0.1 \times (-0.001304) \times (-0.0927) = 0.450012$

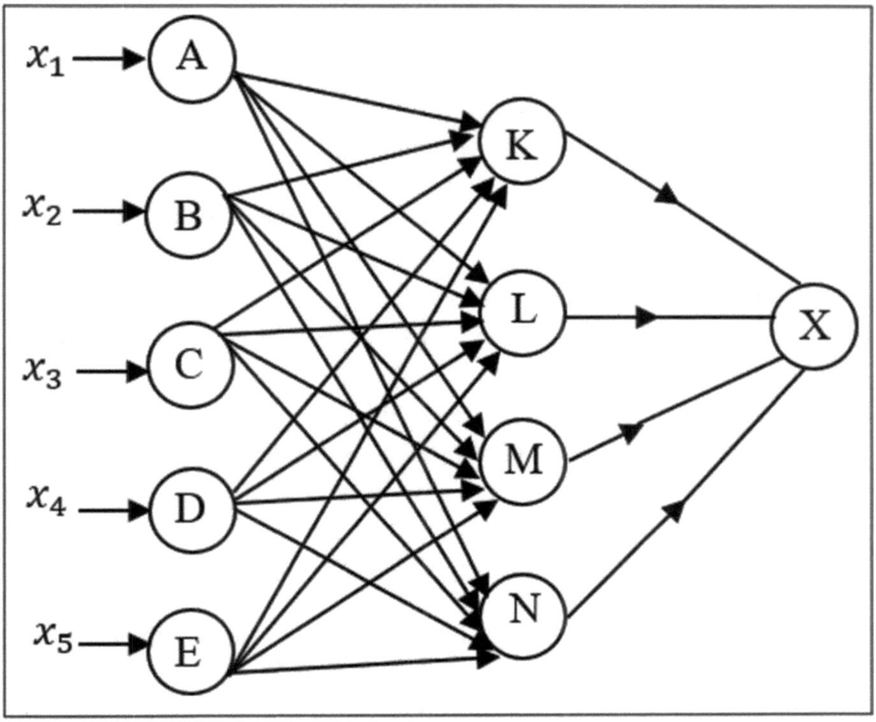

Fig. 3.5 The architecture of ANN for the given problem

value of 1. Initial weights connecting input and hidden nodes are 0.1, whereas these are 0.2 for hidden nodes to output. The learning rate can be considered as 0.2. Use the Hyperbolic Tangent activation function. Establish a relationship using FFBP-ANN and show the computations for one epoch.

Solution:

Input values $x_1, x_2, x_{3,}x_4, x_5$ (Nodes A, B, C, D, E) are 5, 6, 7, 8, 9; Observed value O_X is 1.

Hyperbolic tangent function $f(x) = \frac{e^x - e^{-x}}{e^x + e^{-x}}$

Epoch 1:

Weighted input to the hidden layer $\sum_{i=1}^{5} \omega_i x_i = 0.1 \times 5 + 0.1 \times 6 + 0.1 \times 7 + 0.1 \times 8 + 0.1 \times 9 = 3.5$ [K, L, M, N nodes in the hidden layer]

Output from each node [K, L, M, N] in the hidden layer $O_K = O_L = O_M = O_N = \frac{e^x - e^{-x}}{e^x + e^{-x}} = \frac{e^{3.5} - e^{-3.5}}{e^{3.5} + e^{-3.5}} = 0.9982$

Input to the output layer $= 0.2 \times 0.9982 \times 4 = 0.79856$ [Here 4 represents four hidden nodes that are connected to the output layer]

Predicted output from the output layer X, $P_X = \frac{e^x - e^{-x}}{e^x + e^{-x}} = \frac{e^{0.79856} - e^{-0.79856}}{e^{0.79856} + e^{-0.79856}} = 0.6632$

Observed output $O_X = 1$
Error $E_X = O_X - P_X = 1 - 0.6632 = 0.3368$
The division of errors is $E_{j-1} = O_n(1 - O_n) \sum \omega_{nj} E_j$ [If j is X, $j - 1$ is K, L, M, N]
Entrusting errors to elements in layer 2

$$E_K = O_K \times (1 - O_K) \times (\omega_{KX} \times E_X)$$
$$= 0.9982 \times (1 - 0.9982) \times 0.2 \times 0.3368 = 0.000121$$

$$E_L = O_L \times (1 - O_L) \times (\omega_{LX} \times E_X)$$
$$= 0.9982 \times (1 - 0.9982) \times 0.2 \times 0.3368 = 0.000121$$

$$E_M = O_M \times (1 - O_M) \times (\omega_{MX} \times E_X).$$
$$= 0.9982 \times (1 - 0.9982) \times 0.2 \times 0.3368 = 0.000121$$

$$E_N = O_N \times (1 - O_N) \times (\omega_{NX} \times E_X)$$
$$= 0.9982 \times (1 - 0.9982) \times 0.2 \times 0.3368 = 0.000121$$

Weights are updated as per Eq. 3.3. Tables 3.3 and 3.4 comprise updated weights connecting nodes in layers 1 and 2 & 2 and 3, respectively.

Table 3.3 Updated weights connecting nodes in layers 1 and 2

Updated weight for input to hidden nodes	Updated weights
AK = AL = AM = AN	$0.1 + 0.2 \times 0.000121 \times 5 = 0.100121$
BK = BL = BM = BN	$0.1 + 0.2 \times 0.000121 \times 6 = 0.100145$
CK = CL = CM = CN	$0.1 + 0.2 \times 0.000121 \times 7 = 0.100169$
DK = DL = DM = DN	$0.1 + 0.2 \times 0.000121 \times 8 = 0.100194$
EK = EL = EM = EN	$0.1 + 0.2 \times 0.000121 \times 9 = 0.100218$

Table 3.4 Updated weights connecting nodes in layers 2 and 3

Updated weight for the hidden to output nodes	Updated weights
KX = LX = MX = NX	$0.2 + 0.2 \times 0.000121 \times 0.3368 = 0.200008$

3.4 Wavelet Neural Networks

The principle behind wavelet transform (Guo et al., 2022; Vogl et al., 2022) is to get hold of sufficient insights from primary data and disintegrate the same into further narration. Dilation and translation are essential features in wavelet disintegration. Wavelet transforms are dealt with low and high-pass filters. A standard WNN architecture is presented in Fig. 3.6.

Workflow is as follows: Primary data is established in the input layer, whereas the hidden layer comprises wavelons (or hidden units). The primary input data in the hidden layers are metamorphosed into dilation and translation mother wavelets. Lastly, approximations of the observed values are computed in the output layer (Fig. 3.6). Equation (3.4) denotes the single hidden layer process of feed-forward WNN, which acts like a linear model when no hidden units exist (Vogeti et al., 2022).

$$y_{\text{sim}} = \omega_\alpha + \sum_{j=1}^{\alpha} \omega_j \mathrm{M}(z_{ij}) + \sum_{i=1}^{n} \omega_i x_i \tag{3.4}$$

x_i = Input value, y_{sim} = Simulated output values; α = Number of wavelons; ω_α = Constant ; ω_i and ω_j =weights of input and hidden network; n = Number of inputs; z_{ij} =Normalized input to wavelon; $\mathrm{M}(z_{ij})$ =Intermediate outputs, which are

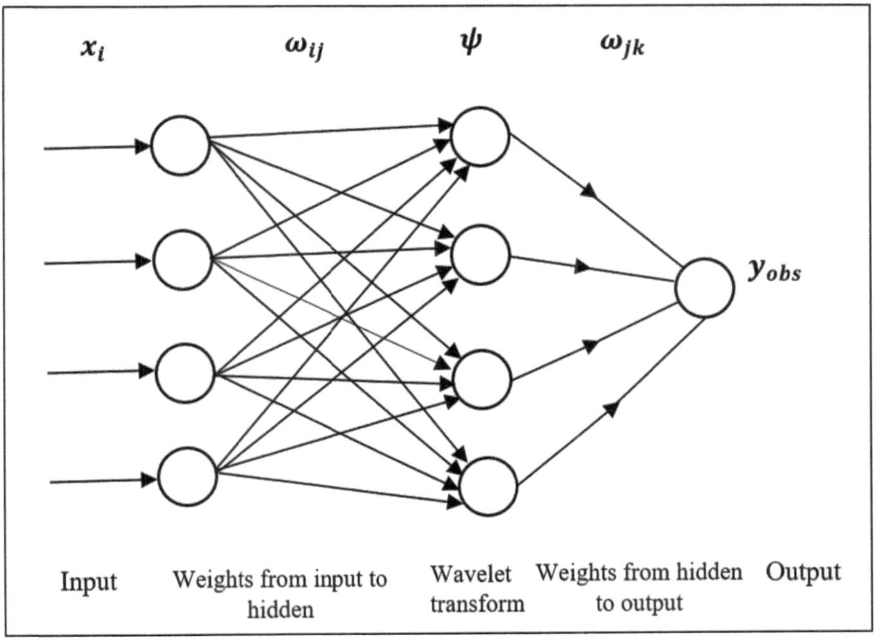

Fig. 3.6 Architecture of WNN

dependent on the selected mother wavelet. They are the crucial components of WNN, which decide the dilation and translation pattern within the Wavelons in the hidden layer. Mathematical expressions of some of the mother wavelets are (Eqs. 3.5–3.8):

$$\text{Gaussian } \mathrm{M}(z_{ij}) = z_{ij} \cdot e^{z_{ij}^2/2} \tag{3.5}$$

$$\text{Mexican hat } \mathrm{M}(z_{ij}) = \left(1 - z_{ij}^2\right) \cdot e^{z_{ij}^2/2} \tag{3.6}$$

$$\text{Morlet } \mathrm{M}(z_{ij}) = e^{-z_{ij}^2/2} . \cos 5\, z_{ij} \tag{3.7}$$

$$\text{Shannon } \mathrm{M}(z_{ij}) = \sin c(t) . e^{-2\pi i t} \tag{3.8}$$

The workflow of WNN is discussed below:

- Identifying the number of hidden units, type of mother wavelet, dilation, and translation and related parameters
- Random assignment of initial network weights
- Computation of y_{sim} and estimating the error
- Updating the weights (and associated parameters) till the chosen termination criteria are fulfilled.

Most of the parameters mentioned above influence network weights significantly, except, Dropout. It purposefully discards wavelons from the network to improve further.

Epochs can be continued till a tolerance error is achieved with adequately assigned learning and momentum rates. Equation 3.9 represents the total error (E_ω), i.e., the difference between observed output (y_{obs}) and the y_{sim}, weighted by the wavelet function (Z) for each training example.

$$E_\omega = \sum (y_{\text{obs}} - y_{\text{sim}}) \times \Psi(Z) \tag{3.9}$$

The error serves as a performance indicator for the network and is typically minimized during training. Repeat the analysis using the new weights and modify the mother wavelets until the observed and simulated values are closer. Once reasonable simulated values are obtained by adjusting the weights and mother wavelets further, the E_ω is back-propagated by using the chain rule till tolerance error is achieved (Yang et al., 2009).

Numerical problem 3.4. Using WNN, establish the relationship between stress (input x) and strain (output y). Datasets are presented in Table 3.5.

Table 3.5 Information about datasets

Dataset	x	y
1	5	0.05
2	9	0.03
3	19	0.2
4	4	0.5
5	21	0.22

Solution:

Step 1: Selection of the number of hidden units, Activation functions (here it is Mother wavelet)
Number of input variables $i = 1$;
Number of output variables $k = 1$;
Number of training records $P = 5$;
The number of hidden units is $n = 1$ (considered only one hidden unit), and the type of mother wavelet function considered is Gaussian wavelet.

Step 2: Assigning $\omega_{\omega 1}$ with a matrix size of $n \times i$ to multiply with input.
$\omega_{\omega 1}$ is assigned based on random weight allocation and matrix size $n \times i$. In this numerical, the matrix size is 1×1
$\omega_{\omega 1} = [0.35]$ (Say the weights are randomly assigned)
Step 3: Computation of weighted input

Weighted input $= \mathrm{I} = 0.35 \times x$
Step 4: Assigning translation (b) and dilation parameter (a)

$$b = 0.5(\mathrm{M_i} + \mathrm{N_i})$$

$$a = 0.2(\mathrm{M_i} - \mathrm{N_i})$$

where,

$$\mathrm{M_i} = \text{Maximum of the input}\, x = 21$$

$$\mathrm{N_i} = \text{Minimum of the input}\, x = 4$$

$$b = 0.5(\mathrm{M_i} + \mathrm{N_i}) = 0.5(21 + 4) = 12.5$$

$$a = 0.2(\mathrm{M_i} - \mathrm{N_i}) = 0.2(21 - 4) = 3.4$$

Step 5: Computation of Z value (u-term)

$$\text{Z value} = \frac{\text{I} - b}{a} = \frac{\text{I} - 12.5}{3.4}$$

Step 6: Computation of hidden unit value or Ψ value

$$\Psi(\text{Z}) = e^{-z^2/2}$$

Step 7: Multiplication of weight matrix $\omega_{\omega 2}$ of size $n \times k$ with a hidden layer. $\omega_{\omega 2}$ is assigned based on random weight allocation, and the matrix size is $n \times k$. In this numerical, the matrix size is 1×1

$\omega_{\omega 2} = [0.45]$ (Say the weights are randomly assigned)

Step 8: Computation of simulated values and resulting anomalies (Table 3.6)

$$y_{\text{sim}} = \omega_{\omega 2}\, \Psi(\text{Z})$$

$$\text{Average total error} = \frac{\sum_{i=1}^{n} \text{E}_i}{n} = 0.1476$$

$$\begin{aligned} \text{E}_\omega &= \sum (y_{\text{obs}} - y_{\text{sim}}) \times \Psi(\text{Z}) \\ &= (0.047) \times (0.0068) + (0.02) \times (0.0228) + (0.097) \\ &\quad \times (0.2278) + (0.498) \times (0.0049) + (0.076) \times (0.3198) \\ &= 0.04962 \end{aligned}$$

Numerical problem 3.5. Establish the relationship between Depth of water, Velocity of flow (x_1 & x_2), and Flood damage (y) using WNN. Information is presented in Table 3.7.

Table 3.6 Computation of simulated strains (y) and anomalies

x	Weighted input = $\text{I} = 0.35 \times x$	Z value = $\frac{\text{I}-b}{a}$	$\Psi(\text{Z}) = e^{-z^2/2}$	$y_{\text{sim}} = \omega_{\omega 2\times}\, \Psi(\text{Z})$	y_{obs}	$\text{E}_i = y_{\text{obs}} - y_{\text{sim}}$
5	1.75	−3.16	0.0068	0.003	0.05	0.047
9	3.15	−2.75	0.0228	0.01	0.03	0.02
19	6.65	−1.72	0.2278	0.103	0.2	0.097
4	1.4	−3.26	0.0049	0.0022	0.5	0.498
21	7.35	−1.51	0.3198	0.144	0.22	0.076

Table 3.7 Information about datasets

Dataset	x_1	x_2	y
1	1	2	0.5
2	4	3	0.7
3	7	4	0.1
4	9	8	0.4
5	11	9	0.6

Solution:

Step 1: Selection of the number of hidden units, Activation functions (here it is Mother Wavelet)
Number of input variables $i = 2$;
Number of output variables $k = 1$;
Number of training records P = 5;
Number of hidden units, $n = 1$ (considered only one hidden unit), and the type of the mother wavelet function considered is Mexican Hat.

Step 2: Assigning $\omega_{\omega 1}$ with a matrix size of $n \times i$ to multiply with input.
$\omega_{\omega 1}$ is assigned based on random weight allocation, and the matrix size $\mathrm{n} \times \mathrm{i}$. In this numerical, the matrix size is 1×2

$\omega_{\omega 1} = [0.3\ 0.4]$ (Say the weights are randomly assigned)

Step 3: Computation of weighted input.
Weighted input = $\mathrm{I} = 0.3 \times x_1 + 0.4 \times x_2$
Step 4: Assigning translation (b) and dilation parameter (a)

$$b = 0.5(\mathrm{M_i} + \mathrm{N_i})$$

$$a = 0.2(\mathrm{M_i} - \mathrm{N_i})$$

where,

$$\mathrm{M_i} = \text{Maximum of the input}\, x_i = 11$$

$$\mathrm{N_i} = \text{Minimum of the input}\, x_i = 1$$

$$b = 0.5(\mathrm{M_i} + \mathrm{N_i}) = 0.5(11 + 1) = 6$$

$$a = 0.2(\mathrm{M_i} - \mathrm{N_i}) = 0.2(11 - 1) = 2$$

Step 5: Computation of Z value (u-term)

$$\text{Z value} = \frac{\text{I} - b}{a} = \frac{\text{I} - 6}{2}$$

Step 6: Computation of hidden unit value or Ψ value

$$\Psi(Z) = \left(1 - z^2\right).e^{-z^2/2}$$

Step 7: Multiplication of weight matrix $\omega_{\omega 2}$ of size $n \times k$ with a hidden layer. $\omega_{\omega 2}$ is assigned based on random weight allocation, and the matrix size $n \times k$. In this numerical, the matrix size is 1×1

$\omega_{\omega 2} = [0.4]$ (Say the weights are randomly assigned)

Step 8: Computation of simulated values and resulting errors (Table 3.8)

$$y_{\text{sim}} = \omega_{\omega 2}\Psi(Z)$$

$$\text{Average total error} = \frac{\sum_{i=1}^{n} \text{E}_i}{n} = 0.552$$

$$\begin{aligned}
\text{E}_\omega &= \sum (y_{\text{obs}} - y_{\text{sim}}) \times \Psi(Z) \\
&= (0.5995) \times (-0.2487) + (0.8773) \times (-0.4433) + (0.1666) \\
&\quad \times (-0.1665) + (0.7985) \times (-0.9963) + (0.3117) \times (0.7207) \\
&= -1.1366
\end{aligned}$$

Table 3.8 Computation of simulated flood damage (y) and error

x_1	x_2	Weighted input $= \text{I} = 0.3 \times x_1 + 0.4 \times x_2$	Z value $= \frac{\text{I}-b}{a}$	$\Psi(Z) = \left(1 - z^2\right).e^{-z^2/2}$	$y_{\text{sim}} = \omega_{\omega 2} \times \Psi(Z)$	y_{obs}	$\text{E}_i = y_{\text{obs}} - y_{\text{sim}}$
1	2	1.1	−2.45	−0.2487	−0.0995	0.5	0.5995
4	3	2.4	−1.8	−0.4433	−0.1773	0.7	0.8773
7	4	3.7	−1.15	−0.1665	−0.0666	0.1	0.1666
9	8	5.9	−0.05	−0.9963	−0.3985	0.4	0.7985
11	9	6.9	0.45	0.7207	0.28828	0.6	0.3117

3.5 Support Vector Regression

The primary focus of SVR is to determine the best hyperplane that suits most datasets by maximizing the error margin (Granata et al., 2016; Raghavendra & Deka, 2014; Vapnik, 1998) (Fig. 3.7). It has a robust theoretical framework that incorporates principles of convex optimization, ML, statistics, and mathematical analysis. These also have good generalization performance, strong adaptability, and the ability to handle non-linearity and noisy datasets compared to traditional algorithms (Madhuri et al., 2021; Mohammadi, 2021; Sujay & Paresh, 2014)). Mathematically, a hyperplane can be denoted (Eq. 3.10):

$$\omega^{\mathrm{T}} x + b = 0 \tag{3.10}$$

where ω^{T} and b are weight vector and bias.

Equation 3.11 defines the decision boundary of the hyperplane, $\omega^{\mathrm{T}} x + b = 0$. If $\omega^{\mathrm{T}} x + b > 0$, then the datasets related to overestimated vectors; if $\omega^{\mathrm{T}} x + b < 0$, the dataset related to underestimated vectors. The remaining vectors that fall within the error margin are termed support vectors. The optimal hyperplane can be identified by satisfying the constraint of maximizing the error margin by minimizing the objective function (Eq. 3.11)

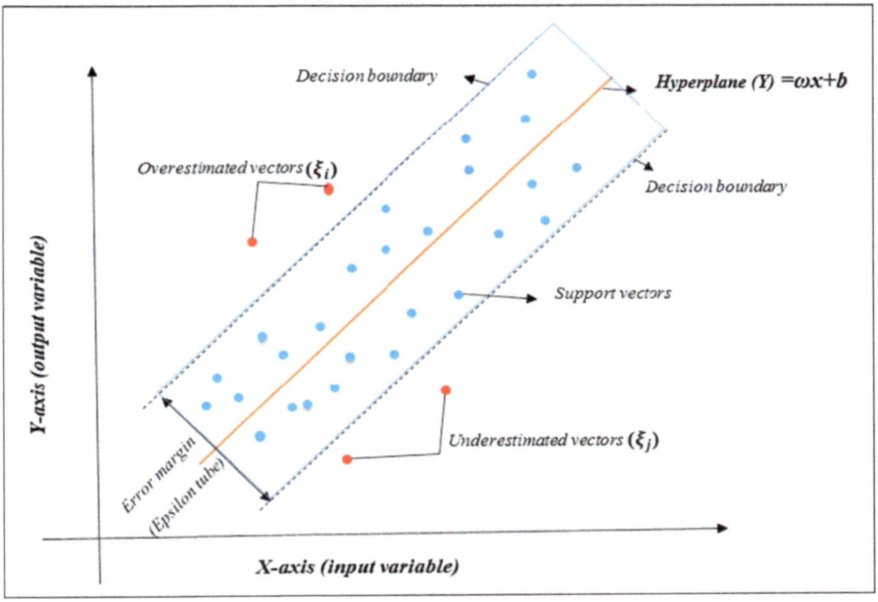

Fig. 3.7 Hyperplane of SVR

$$Z = \frac{1}{2}.|\omega|^2 \tag{3.11}$$

subject to $y_i(\omega^T x_i + b) \geq 1$ for $i = 1, 2, ..., n$, where $|\omega|^2$ is the squared normalized weight vector ω; x_i and y_i are ith datasets in the feature space. The constraint $y_i(\omega^T x_i + b) \geq 1$ ensures the margin is at least 1 for all datasets. The optimization problem can be worked out utilizing the Lagrange multiplier approach, and the dual form of the problem is defined as the (Eq. 3.12)

$$\text{Maximize L} = \text{sum}(\alpha_i) - \left(\frac{1}{2}\right) \times \text{sum}(\text{sum}(\alpha_i.\alpha_j \cdot y_i \cdot y_j \cdot x_i \cdot x_j)) \tag{3.12}$$

subject to $\alpha_i \geq 0$ and sum $(\alpha_i, y_i) = 0$

where α_i is the Lagrange multiplier for the ith constraint. The decision boundary can be found using Eq. 3.13.

$$f(x) = \text{sign}(\text{sum}(\alpha_i \cdot y_i \cdot x_i \cdot x_j) + b) \tag{3.13}$$

It can be extended to non-linearly separable datasets using the kernel trick. Some kernels are the Radial Basis Function (RBF), Sigmoid, and polynomial. The mathematical formula for the RBF kernel is (Eq. 3.14):

$$K(x_i, x_j) = e^{\frac{-|x_i - x_j|^2}{2\sigma^2}} \tag{3.14}$$

where σ is a hyperparameter that controls the width of the RBF function, the parameters governing SVR are the kernel, shape, regularization parameters, and dropout. The loss function can be computed between observed and predicted values. Local gradients can be updated until the termination criteria are satisfied (Madhuri, 2022).

Numerical problem 3.6. Using SVR, relate rainfall (x) and drought index (y). Datasets are presented in Table 3.9. Compute output value for input of 6. Assume suitable data, if any.

Table 3.9 Information about input and output

Dataset	x	y
1	2	7
2	5	37
3	4	17

Solution:

The given problem is solved in a stepwise manner, as presented below

Step 1: Calculation of kernel matrix K by selecting the appropriate kernel function

The computation of kernel elements (k_{ij}) depends on the chosen kernel function. Here, RBF is selected as the kernel; thus, the elements of the kernel present in the matrix are computed based on the RBF expression as $k_{ij} = e^{-\Upsilon \|x_i - x_j\|^2}$. Here, Υ is a positive constant controlling the spread of a kernel chosen as 0.1 in this problem. The choice of 0.1 as a default value for Υ is often a reasonable starting point for a wide range of datasets. It is neither too small nor too large, allowing for a balance between capturing local patterns and generalizing well to unseen datasets.

$x_1 = 2, x_2 = 5$, and $x_3 = 4$

Kernel Elements can be represented as follows:

	x_1	x_2	x_3
x_1	k_{11}	k_{12}	k_{13}
x_2	k_{21}	k_{22}	k_{23}
x_3	k_{31}	k_{32}	k_{33}

Calculations of the kernel elements based on the input elements are presented as follows (Table 3.10)

$$k_{11} = e^{\left(-\Upsilon\|x_1-x_1\|^2\right)} = e^{\left(-\Upsilon\|2-2\|^2\right)} = e^0 = 1$$
$$k_{12} = e^{\left(-\Upsilon\|x_1-x_2\|^2\right)} = e^{\left(-\Upsilon\|2-5\|^2\right)} = e^{-0.1\times 9} = 0.4065$$
$$k_{13} = e^{\left(-\Upsilon\|x_1-x_3\|^2\right)} = e^{\left(-\Upsilon\|2-4\|^2\right)} = e^{-0.1\times 4} = 0.6703$$

$$k_{21} = e^{\left(-\Upsilon\|x_2-x_1\|^2\right)} = e^{\left(-\Upsilon\|5-2\|^2\right)} = e^{-0.1\times 9} = 0.4065$$
$$k_{22} = e^{\left(-\Upsilon\|x_2-x_2\|^2\right)} = e^{\left(-\Upsilon\|5-5\|^2\right)} = e^0 = 1$$
$$k_{23} = e^{\left(-\Upsilon\|x_2-x_3\|^2\right)} = e^{\left(-\Upsilon\|5-4\|^2\right)} = e^{-0.1\times 1} = 0.9048$$

$$k_{31} = e^{\left(-\Upsilon\|x_3-x_1\|^2\right)} = e^{\left(-\Upsilon\|4-2\|^2\right)} = e^{-0.1\times 4} = 0.6703$$
$$k_{32} = e^{\left(-\Upsilon\|x_3-x_2\|^2\right)} = e^{\left(-\Upsilon\|4-5\|^2\right)} = e^{-0.1\times 1} = 0.9048$$
$$k_{33} = e^{\left(-\Upsilon\|x_3-x_3\|^2\right)} = e^{\left(-\Upsilon\|4-4\|^2\right)} = e^0 = 1$$

Step 2: Computation of contraction coefficient (α) based on observed variables and the obtained kernel matrix (K), which is as follows:

The output y is the function of kernel matrix (K) and contraction coefficient (α)

$$y = K \cdot \alpha$$

Table 3.10 Kernel matrix $K =$

	x_1	x_2	x_3
x_1	1	0.4065	0.6703
x_2	0.4065	1	0.9048
x_3	0.6703	0.9048	1

(or)

$$\alpha = (\mathrm{K})^{-1} \cdot y$$

where, $y = \begin{bmatrix} 7 \\ 37 \\ 17 \end{bmatrix}$

In this numerical problem, the Gauss-Jordan elimination procedure is used. The augmented matrix of the K is as follows =

1	0.4065	0.6703	**1**	**0**	**0**
0.4065	1	0.9048	**0**	**1**	**0**
0.6703	0.9048	1	**0**	**0**	**1**

Sequentially pivot the diagonal elements for the first three columns and apply row operations such that diagonal elements are one and the remaining elements are zeroes. The row operations applied to the matrix are as follows:

$\mathrm{R}_2 = \mathrm{R}_2 - 0.4065\mathrm{R}_1$; $\mathrm{R}_3 = \mathrm{R}_3 - 0.6703\mathrm{R}_1$; $\mathrm{R}_2 = 1.1979\mathrm{R}_2$; $\mathrm{R}_3 = \mathrm{R}_3 - 0.6703\mathrm{R}_1$; $\mathrm{R}_3 = 13.9435\mathrm{R}_3$; $\mathrm{R}_1 = \mathrm{R}_1 - 0.3623\,\mathrm{R}_3$ and $\mathrm{R}_2 = \mathrm{R}_2 - 0.7575\mathrm{R}_3$.

The inverse matrix computed from Gauss-Jordan elimination is as follows: $(\mathrm{K})^{-1} =$

3.029	3.3405	−5.0528
3.3405	9.1987	−10.5621
−5.0528	−10.5621	13.9435

$$(\mathrm{K})^{-1} \cdot \mathrm{y} = \alpha$$

$(K)^{-1}$			y		α
3.029	3.3405	−5.0528	7	=	58.904
3.3405	9.1987	−10.5621	37		184.18
−5.0528	-10.5621	13.9435	17		−189.128

$$C_{11} = 3.029 \times 7 + 3.3405 \times 37 + (-5.0528) \times 17 = 58.904$$
$$C_{21} = 3.3405 \times 7 + 9.1987 \times 37 + (-10.5621) \times 17 = 184.18$$
$$C_{31} = (-5.0528) \times 7 + (-10.5621) \times 37 + 13.9435 \times 17 = -189.128$$

Step 3: Computation of bias (b)

$$b_k = y_k - \sum \alpha_i k_{pj}\,;\ p = 1$$
$$b_1 = y_1 - \sum \alpha_i k_{1j}$$
$$b_1 = 7 - [(58.904 \times 1) + (184.18 \times 0.4065) + (-189.128 \times 0.6703)]$$
$$b_1 = -0.0006716$$

$$b_k = y_k - \sum \alpha_i k_{pj}\,;\ p=2$$
$$b_2 = y_2 - \sum \alpha_i k_{2j}$$
$$b_2 = 37 - [(58.904 \times 0.4065) + (184.18 \times 1) + (-189.128 \times 0.9048)]$$
$$b_2 = -0.0014616$$

$$b_k = y_k - \sum \alpha_i k_{pj};\ p = 3$$
$$b_3 = y_3 - \sum \alpha_i k_{3j}$$
$$b_3 = 17 - [(58.904 \times 0.6703) + (184.18 \times 0.9048) + (-189.128 \times 1)]$$
$$b_3 = -0.0014152$$

$$b = \frac{b_1 + b_2 + b_3}{3}$$
$$b = \frac{(-0.0006716) + (-0.0014616) + (-0.0014152)}{3}$$
$$b = -0.0011828$$

Step 4: Computation of simulated values using density function $f(x_i)$
Density function $f(x_i) = \sum \alpha_m k_{ij} + b\,(m=\{1, 2, 3\},\ j = \{1, 2, 3\})$

For $i = 1, f(x_1) = \sum \alpha_m k_{1j} + b = \alpha_1 k_{11} + \alpha_2 k_{12} + \alpha_3 k_{13} + b$

$$f(x_1) = (58.904 \times 1) + (184.18 \times 0.4065) + (-189.128 \times 0.6703) + (-0.0011828)$$
$$f(x_1) = 6.99948$$

For $i = 2, f(x_2) = \sum \alpha_m k_{2j} + b = \alpha_1 k_{21} + \alpha_2 k_{22} + \alpha_3 k_{23} + b$

$$f(x_2) = (58.904 \times 0.4065) + (184.18 \times 1) + (-189.128 \times 0.9048) + (-0.0011828)$$
$$f(x_2) = 37.0002$$

For $i = 3, f(x_3) = \sum \alpha_m k_{3j} + b = \alpha_1 k_{31} + \alpha_2 k_{32} + \alpha_3 k_{33} + b$

$$f(x_3) = (58.904 \times 0.6703) + (184.18 \times 0.9048) + (-189.128 \times 1) + (-0.0011828)$$
$$f(x_3) = 16.8482.$$

Step 5: Computation of loss function between observed and simulated values (Table 3.11).

The simulated values have a good agreement with the observed values; thus, in this numerical, the density function can be used for predicting the new input values (x_n) as presented in Step 6. If the simulated values deviate significantly from the observed values, the kernel functions can be changed (for example, Linear, Polynomial, Gaussian, Logistic).

Step 6: Computation of kernel and output using density function $f(x_i)$ for the new input value x_n

The kernel calculated for input, $x_n = 6$ are presented as follows:

$$k(x_i, x_n) = e^{-\Upsilon \|x_i - x_j\|^2}$$

For $i = 1, k(x_1, x_n) = e^{-\Upsilon \|x_1 - x_n\|^2} = e^{-0.1\|2-6\|^2} = 0.2019$
For $i = 2, k(x_2, x_n) = e^{-\Upsilon \|x_2 - x_n\|^2} = e^{-0.1\|5-6\|^2} = 0.9048.$
For $i = 3, k(x_3, x_n) = e^{-\Upsilon \|x_3 - x_n\|^2} = e^{-0.1\|4-6\|^2} = 0.6703.$

Table 3.11 Observed, simulated, and loss function values

Dataset	Observed values (y_k)	Simulated values $f(x_n)$	Loss function $L = y_k - f(x_n)$
1	7	6.99948	0.00052
2	37	37.0002	−0.0002
3	17	16.8482	0.1518

Now, applying the updated kernel elements in the density function, $f(x_n) = \sum \alpha_i k_{nj} + b$

$$f(x_n) = (58.904 \times 0.2019) + (184.18 \times 0.9048) + (-189.128 \times 0.6703) + (-0.0011828)$$
$$= 51.765.$$

For the new value (x_n), output value predicted using the density function is $f(x_n) = 51.765$.

Numerical problem 3.7. Relate wheel load (x) and pavement failure (y) using the SVR for the datasets presented in Table 3.12. What is the output for an input value of 6?

Solution:

The given problem is solved in a stepwise manner, as presented below:
Step 1: Calculation of kernel matrix K by selecting the appropriate kernel function
The computation of kernel elements (k_{ij}) depends on the chosen kernel function. Here, RBF is selected as the kernel; thus, the elements of the kernel present in the matrix are computed based on the RBF expression *as* $k_{ij} = e^{-\Upsilon \|x_i - x_j\|^2}$. Here, Υ was chosen as 0.1 in this problem. $x_1 = 3, x_2 = 7, x_3 = 5$, and $x_4 = 14$.
Kernel Elements

	x_1	x_2	x_3	x_4
x_1	k_{11}	k_{12}	k_{13}	k_{14}
x_2	k_{21}	k_{22}	k_{23}	k_{24}
x_3	k_{31}	k_{32}	k_{33}	k_{34}
x_4	k_{41}	k_{41}	k_{43}	k_{44}

Calculation of the kernel elements based on the input elements are as follows: (Table 3.13)

$$k_{11} = e^{(-\gamma \|x_1 - x_1\|^2)} = e^{(-\gamma \|3-3\|^2)} = e^0 = 1$$
$$k_{12} = e^{(-\Upsilon \|x_1 - x_2\|^2)} = e^{(-\Upsilon \|3-7\|^2)} = e^{-0.1\times 16} = 0.2019.$$
$$k_{13} = e^{(-\Upsilon \|x_1 - x_3\|^2)} = e^{(-\Upsilon \|3-5\|^2)} = e^{-0.1\times 4} = 0.6703.$$

Table 3.12 Information about input and output

Dataset	x	y
1	3	1.2
2	7	3
3	5	1.6
4	14	6.5

$$k_{14} = e^{(-\gamma\|x_1 - x_4\|^2)} = e^{(-\gamma\|3-14\|^2)} = e^{(-0.1\times121)} = 0$$

$$k_{21} = e^{(-\Upsilon\|x_2 - x_1\|^2)} = e^{(-\Upsilon\|7-3\|^2)} = e^{-0.1\times16} = 0.2019.$$
$$k_{22} = e^{(-\Upsilon\|x_2 - x_2\|^2)} = e^{(-\Upsilon\|7-7\|^2)} = e^{0} = 1.$$
$$k_{23} = e^{(-\Upsilon\|x_2 - x_3\|^2)} = e^{(-\Upsilon\|7-5\|^2)} = e^{-0.1\times4} = 0.6703.$$
$$k_{24} = e^{(-\Upsilon\|x_2 - x_4\|^2)} = e^{(-\Upsilon\|7-14\|^2)} = e^{-0.1\times49} = 0.00745$$

$$k_{31} = e^{(-\Upsilon\|x_3 - x_1\|^2)} = e^{(-\Upsilon\|5-3\|^2)} e^{-0.1\times4} = 0.6703.$$
$$k_{32} = e^{(-\Upsilon\|x_3 - x_2\|^2)} = e^{(-\Upsilon\|5-7\|^2)} = e^{-0.1\times4} = 0.6703.$$
$$k_{33} = e^{(-\Upsilon\|x_3 - x_3\|^2)} = e^{(-\Upsilon\|5-5\|^2)} = e^{0} = 1.$$
$$k_{34} = e^{(-\Upsilon\|x_3 - x_4\|^2)} = e^{(-\Upsilon\|5-14\|^2)} = e^{-0.1\times81} = 0.0003.$$

$$k_{41} = e^{(-\Upsilon\|x_4 - x_1\|^2)} = e^{(-\Upsilon\|14-3\|^2)} = e^{-0.1\times121} = 0.$$
$$k_{42} = e^{(-\Upsilon\|x_4 - x_2\|^2)} = e^{(-\Upsilon\|14-7\|^2)} = e^{-0.1\times49} = 0.00745.$$
$$k_{43} = e^{(-\Upsilon\|x_4 - x_3\|^2)} = e^{(-\Upsilon\|14-5\|^2)} = e^{-0.1\times81} = 0.0003.$$
$$k_{44} = e^{(-\Upsilon\|x_4 - x_4\|^2)} = e^{(-\Upsilon\|14-14\|^2)} = e^{0} = 1.$$

Step 2: Computation of contraction coefficient (α) based on observed variables and kernel matrix (K) obtained

The output y is the function of kernel matrix (K) and contraction coefficient (α)

$$y = K \cdot \alpha$$

or

$$\alpha = (\mathrm{K})^{-1} \cdot y$$

Table 3.13 Kernel matrix K =

	x_1	x_2	x_3	x_4
x_1	1	0.2019	0.6703	0
x_2	0.2019	1	0.6703	0.00745
x_3	0.6703	0.6703	1	0.0003
x_4	0	0.00745	0	1

where, $y = \begin{bmatrix} 1.2 \\ 3 \\ 1.6 \\ 6.5 \end{bmatrix}$

The augmented matrix of the K is presented as follows.

1	0.2019	0.6703	0	**1**	**0**	**0**	**0**	**0**
0.2019	1	0.6703	0.00745	**0**	**1**	**0**	**0**	**0**
0.6703	0.6703	1	0.0003	**0**	**0**	**1**	**0**	**0**
0	0.00745	0	1	**0**	**0**	**0**	**1**	**0**

Sequentially pivot the diagonal elements for the first five columns and apply row operations such that diagonal elements are one and the remaining elements are zeroes. The row operations applied to the matrix are as follows (as part of the Gauss-Jordan Elimination procedure)

$R_2 = R_2 - 0.2019R_1; R_3 = R_3 - 0.6703R_1;\ R_2 = 1.0425R_2; R_3 = R_3 - 0.534966R_2;$ $R_4 = R_4 - 0.00745R_2;\ R_3 = 3.9628R_3;\ R_4 = R_4 + 0.004155R_3;\ R_4 = 1.00012R_4;$ $R_3 = R_3 + 0.015276R_4;\ R_2 = R_2 - 0.007767R_4;\ R_2 = R_2 - 0.5577R_3;\ R_1 = R_1 -$ $0.6703R_3;\ R_1 = R_1 - 0.2019R_2.$

The inverse matrix computed from Gauss-Jordan elimination is as follows: $(K)^{-1} =$

2.2751	1.0222	−2.2102	−0.00695
1.02219	2.2753	−2.2103	−0.01629
2.2102	−2.2103	3.9630	0.01528
−0.007615	−0.01695	0.01646	1.0001

$$\alpha = \begin{bmatrix} 2.2152 \\ 4.4102 \\ 2.4615 \\ 6.467 \end{bmatrix}$$

$C_{11} = 2.2751 \times 1.2 + 1.0222 \times 3 + (-2.2102) \times 1.6 + (-0.00695) \times 6.5 = 2.2152.$

$C_{21} = 1.02219 \times 1.2 + 2.2753 \times 3 + (-2.2103) \times 1.6 + (-0.01629) \times 6.5 = 4.4102.$

$C_{31} = 2.2102 \times 1.2 + (-2.2103) \times 3 + 3.9630 \times 1.6 + 0.01528 \times 6.5 = 2.4615.$

$C_{41} = (-0.007615) \times 1.2 + (-0.01695) \times 3 + 0.01646 \times 1.6 + 1.0001 \times 6.5 = 6.467.$

Step 3: Computation of Bias (b)

$$b_k = y_k - \sum \alpha_i k_{pj},\ p = 1$$
$$b_1 = y_1 - \sum \alpha_i k_{1j}$$

$b_1 = 1.2 - [(2.2152 \times 1) + (4.4102 \times 0.2019) + (2.4615 \times 0.6703) + (6.467 \times 0)]$
$b_1 = -3.555$

$$b_k = y_k - \sum \alpha_i k_{pj};\ p = 2$$

$$b_2 = y_2 - \sum \alpha_i k_{2j}$$

$$b_2 = 3 - [(2.2152 \times 0.2019) + (4.4102 \times 1) + (2.4615 \times 0.6703) + (6.467 \times 0.00745)]$$

$$b_2 = -3.555$$

$$b_k = y_k - \sum \alpha_i k_{pj}; p = 3$$
$$b_3 = y_3 - \sum \alpha_i k_{3j}$$

$b_3 = 1.6 - [(2.2152 \times 0.6703) + (4.4102 \times 0.6703) + (2.4615 \times 1) + (6.467 \times 0.0003)]$
$b_3 = -5.3044$

$$b_k = y_k - \sum \alpha_i k_{pj};\ p = 4$$

$b_4 = y_4 - \sum \alpha_i k_{4j}$
$b_4 = 6.5 - [(2.2152 \times 0) + (4.4102 \times 0.00745) + (2.4615 \times 0) + (6.467 \times 1)]$
$b_4 = -0.000144$

$$b = \frac{b_1 + b_2 + b_3 + b_4}{4}$$

$$b = \frac{(-3.555) + (-3.555) + (-5.3044) + (-0.000144)}{4}$$

$$b = -3.1036$$

Step 4: Computation of simulated values using Density Function $f(x_i)$

Density function $f(x_i) = \sum \alpha_m k_{ij} + b(m = \{1, 2, 3, 4\},\ j = \{1, 2, 3, 4\})$
For $i = 1, f(x_1) = \sum \alpha_1 k_{1j} + b = \alpha_1 k_{11} + \alpha_2 k_{12} + \alpha_3 k_{13} + \alpha_4 k_{14} + b$

$$f(x_1) = (2.2152 \times 1) + (4.4102 \times 0.2019) + (2.4615 \times 0.6703) + (6.647 \times 0) + (-3.1036)$$

$f(x_1) = 1.65.$

For $i = 2, f(x_2) = \sum \alpha_2 k_{2j} + b = \alpha_1 k_{21} + \alpha_2 k_{22} + \alpha_3 k_{23} + \alpha_4 k_{24} + b$

$$f(x_2) = (2.2152 \times 0.2019) + (4.4102 \times 1) + (2.4615 \times 0.6703) + (6.647 \times 0.00745) + (-3.1036)$$
$$f(x_2) = 3.4533$$

For $i = 3, f(x_3) = \sum \alpha_3 k_{3j} + b = \alpha_1 k_{31} + \alpha_2 k_{32} + \alpha_3 k_{33} + \alpha_4 k_{34} + b.$

$$f(x_3) = (2.2152 \times 0.6703) + (4.4102 \times 0.6703) + (2.4615 \times 1) + (6.647 \times 0.0003) + (-3.1036)$$
$$f(x_3) = 3.8009.$$

For $i = 4, f(x_4) = \sum \alpha_4 k_{4j} + b = \alpha_1 k_{41} + \alpha_2 k_{42} + \alpha_3 k_{43} + \alpha_4 k_{44} + b.$

$$f(x_4) = (2.2152 \times 0) + (4.4102 \times 0.00745) + (2.4615 \times 0) + (6.647 \times 1) + (-3.1036)$$
$$f(x_4) = 3.5762.$$

Step 5: Computation of loss function between observed and simulated values (Table 3.14)

The simulated values have a moderate agreement with the observed values and require further updation of the kernel functions, contraction coefficients, and biases until the deviation is minimized to the best possible extent. However, in this numerical, the density function is used to predict the new input values (x_n) as presented in Step 6 for demonstration purposes.

Step 6: Computation of kernel and output using density function $f(x_i)$ for the new input value x_n.

The kernel calculated for $x_n = 6$ are presented as follows:

$$k(x_i, x_n) = e^{(-\Upsilon\|x_i - x_n\|^2)}$$

For $i = 1, k(x_1, x_n) = e^{-\Upsilon\|x_1 - x_6\|^2} = e^{-0.1\|3-6\|^2} = 0.4065$
For $i = 2, k(x_2, x_n) = e^{-\Upsilon\|x_2 - x_6\|^2} = e^{-0.1\|7-6\|^2} = 0.9048$
For $i = 3, k(x_3, x_n) = e^{-\Upsilon\|x_3 - x_6\|^2} = e^{-0.1\|5-6\|^2} = 0.9048$

Table 3.14 Observed, simulated, and loss function values

Dataset	Observed values (y_k)	Simulated values $f(x_n)$	Loss function $L = y_k - f(x_n)$
1	1.2	1.65	−0.45
2	3	3.4533	−0.4533
3	1.6	3.8009	−2.2009
4	6.5	3.5762	2.9238

For $i = 4$, $k(x_4, x_n) = e^{-\Upsilon\|x_4 - x_6\|^2} = e^{-0.1\|4-6\|^2} = 0.00166$.

Now, applying the updated kernel elements in the density function, $f(x_n) = \sum \alpha_i k_{nj} + b$

$$f(x_n) = (2.2152 \times 0.4065) + (4.4102 \times 0.9048) + (2.4615 \times 0.9048) + (6.467 \times 0.00166) + (-3.1036) = 4.025$$

For the new value (x_n), output value predicted using the density function is $f(x_n) = 4.025$.

3.6 Extreme Learning Machine

ELM employs only a single-hidden layer in the feed-forward networks framework (Wang et al., 2022). Figure 3.8 presents the architecture. Weights connecting the input and hidden layer are randomly given. Similarly, biases in the hidden layer are randomly assigned. They are kept constant during the training procedure. The algorithm converges faster and is likely to reach a globally optimal solution than several traditional algorithms, as no iterative process is involved during the learning process (Huang et al., 2015). Here, only the basic version of ELM is presented.

The input layer is comprised of multiple nodes, each representing a data feature. The hidden layer is formed by multiplying the input values with a randomly generated weight matrix to create a linear combination. The output layer is a linear combination of the hidden layer with a weight matrix $\beth$ (that connects the hidden and output layers). The training intends to estimate the $\beth_i$. The mathematical formulation is as follows (Eq. 3.15):

$$f_{L(x)} = \sum_{i=1}^{L} \beth i \ \Psi_i(X) = \sum_{i=1}^{L} \beth i \times \Psi_i(\omega_i x_j + b_i), j = 1, 2, \ldots .N \tag{3.15}$$

Here, L and N are the hidden units and training datasets (in numbers), x_n is the input of the nth feature of the dataset; ω_i, b_i are the weight and bias vectors connecting the input and hidden layer $\beth_i$, is the weight vector between ith hidden layer ($i = 1$, 2, L) and output. Ψ is an activation function employed elementwise to the result of the linear transformation.

The process is similar to the back-propagation in standard neural networks and is as follows (Eqs. 3.16–3.17):

$$y_{obs} = \lambda\beth \tag{3.16}$$

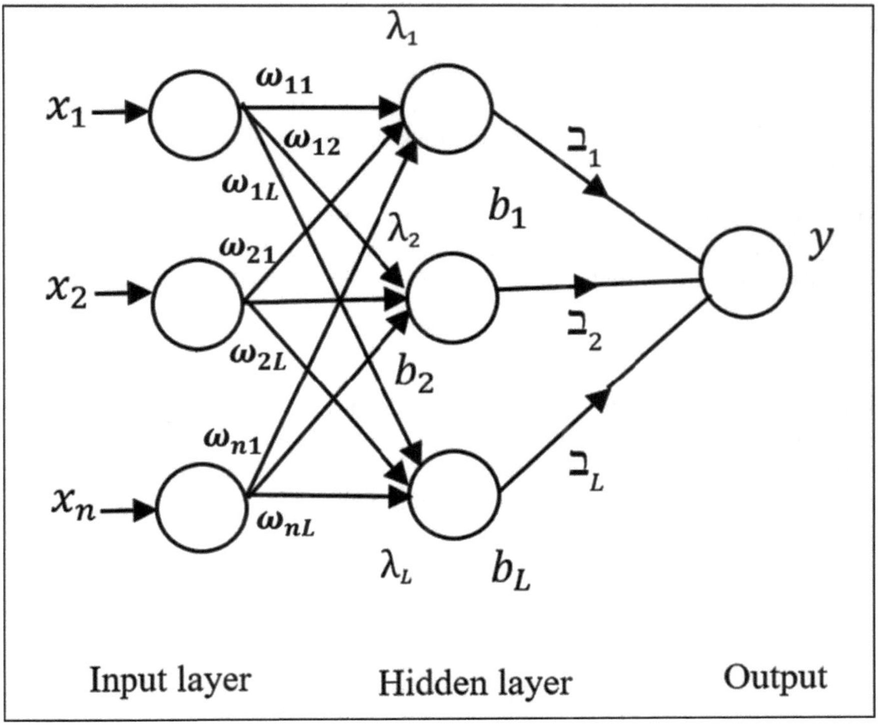

Fig. 3.8 The architecture of basic ELM

$$
\lambda = \begin{bmatrix} \Psi(\omega_1 \times x_1 + b_1) & \cdots & \Psi(\omega_L \times x_1 + b_L) \\ \vdots & \cdots & \vdots \\ \Psi(\omega_1 \times x_N + b_1) & \ldots & \Psi(\omega_L \times x_N + b_L) \end{bmatrix}_{N\times L} \quad \beth = \begin{bmatrix} \beth_1^T \\ \vdots \\ \beth_L^T \end{bmatrix}_{L\times m}
$$

$$
y_{obs} = \begin{bmatrix} y_1^T \\ \vdots \\ y_N^T \end{bmatrix}_{N\times m} \tag{3.17}
$$

λ refers to the hidden output layer matrix linking the input and hidden layer, and m is the number of outputs. Y_{obs} represents the observed data. The working mechanism of ELM is as follows:

- Initial assignment of weights and biases
- Computation of λ
- Computation of the output vector between the hidden and output layer

$\hat{\beth} = \lambda^{\dagger} y_{obs}$; $\lambda^{\dagger}$ refers to Moore–Penrose generalized inverse of the matrix λ, $\hat{\beth}$ is the estimated output weight vector.

- Use $\widehat{\beth}$ to predict new output, $y_{pred} = \lambda\widehat{\beth}$, y_{pred} is predicted output.

It is challenging to understand the working of the model with randomly generated weights. Room for flexibility is minimal as it is a one-layer network. In addition, the algorithm requires careful input scaling to ensure that they are in a range appropriate for the activation function. Otherwise, it may lead to poor prediction accuracy.

Numerical problem 3.8. Table 3.15 presents datasets with two input variables (x_1, x_2), Corrected speed, Pressure ratio, and one output variable (y_{obs}), Corrected flow rate. Analyze the problem in the ELM framework using the Hyperbolic Tangent Function.

Solution:

Step 1: Weights connecting the input and hidden layers are randomly initialized. A set of weights characterizes each node (ω_1, ω_2) and a bias (b). For this example, one hidden layer with two nodes is considered. Two weights are randomly generated for each input feature, which are as follows (Table 3.16):

Step 2: Computation of λ (refer to Eq. 3.17):

$$\lambda = \begin{bmatrix} \Psi(\omega_1 \times x_L + b_1) & \cdots & \Psi(\omega_L \times x_1 + b_L) \\ \vdots & \cdots & \vdots \\ \Psi(\omega_1 \times x_N + b_1) & \ldots & \Psi(\omega_L \times x_N + b_L) \end{bmatrix}_{N \times L}$$

The activation function is Hyperbolic tangent, $\Psi(Z) = \frac{(e^z - e^{-z})}{(e^z + e^{-z})}$.

Table 3.15 Information about input and output

Dataset	x_1	x_2	y_{obs}
1	2	3	4
2	6	11	7

Table 3.16 Random weight matrix

Input Node	Hidden node λ_1	Hidden node λ_2
x_1	0.15	0.35
x_2	0.25	0.50
b	0.1	0.2

Similarly, the below feature matrix λ is obtained by calculating all hidden layer outputs.

$$\lambda = \begin{bmatrix} \Psi(2 \times 0.15 + 3 \times 0.25 + 0.1) & \Psi(2 \times 0.35 + 3 \times 0.5 + 0.2) \\ \Psi(6 \times 0.15 + 11 \times 0.25 + 0.1) & \Psi(6 \times 0.35 + 11 \times 0.5 + 0.2) \end{bmatrix}_{N \times L}$$

$$\lambda = \begin{bmatrix} \Psi(1.15) & \Psi(2.4) \\ \Psi(3.75) & \Psi(7.8) \end{bmatrix}$$

$$\lambda_1 \qquad \lambda_2$$

$$\lambda = \begin{bmatrix} 0.8178 & 0.9836 \\ 0.9989 & 0.9999 \end{bmatrix}$$

Step 3: Computation of output weight vector

$$\hat{\beth} = \lambda^{\dagger} y_{obs}$$

where $\lambda^{\dagger} = (\lambda^{T}\lambda)^{-1}\lambda^{T}$; Here, λ^{T} is the transpose of feature matrix λ, y_{obs} is a vector of observed output. To solve for $\beth$, use the Moore–Penrose pseudoinverse of λ, which is denoted by $\lambda^{\dagger}$:

$$\lambda^{T} = \begin{bmatrix} 0.8178 & 0.9989 \\ 0.9836 & 0.9999 \end{bmatrix}$$

$$(\lambda^{T}\lambda) = \begin{bmatrix} 1.6666 & 1.8032 \\ 1.8032 & 1.9673 \end{bmatrix}$$

$$(\lambda^{T}\lambda)^{-1} = \frac{1}{\det(\lambda^{T}\lambda)} \text{Adjoint}(\lambda^{T}\lambda)$$

$$(\lambda^{T}\lambda)^{-1} = \begin{bmatrix} 72.435 & -66.394 \\ -66.394 & 61.365 \end{bmatrix}$$

$$\lambda^{\dagger} = (\lambda^{T}\lambda)^{-1}\lambda^{T} = \begin{bmatrix} 72.435 & -66.394 \\ -66.394 & 61.365 \end{bmatrix} \begin{bmatrix} 0.8178 & 0.9989 \\ 0.9836 & 0.9999 \end{bmatrix} = \begin{bmatrix} -6.0674 & 5.9685 \\ 6.0613 & -4.9624 \end{bmatrix}$$

$$\text{Output weight vector } \hat{\beth} = \lambda^{\dagger} y_{obs} = \begin{bmatrix} -6.0674 & 5.9685 \\ 6.0613 & -4.9624 \end{bmatrix} \begin{bmatrix} 4 \\ 7 \end{bmatrix}$$

$$\begin{bmatrix} (-6.0674 \times 4) + (5.9685 \times 7) \\ (6.0613 \times 4) + (-4.9624 \times 7) \end{bmatrix} = \begin{bmatrix} 17.5099 \\ -10.4916 \end{bmatrix}$$

The output variable is a linear combination of hidden layer nodes, and the corresponding Equation is

$$\hat{y}_{pred} = (17.5099 \times \lambda_1) + (-10.4916 \times \lambda_2)$$

Using the λ and output variable from the previous steps, predicting the output for the training dataset (by multiplying the $\beth$ with hidden layer output values of the input):

$$\hat{y}_1 = (17.5099 \times 0.8178) + (-10.4916 \times 0.9836) = 4$$
$$\hat{y}_2 = (17.5099 \times 0.9989) + (-10.4916 \times 0.9999) = 7$$

The Mean Square Loss Function (MSLF) is presented in Table 3.17.
MSLF = 0. Hence, the parameters employed are logical.

Numerical problem 3.9. Three-input variables (x_1, x_2, x_3), Catalyst loading, Gasification temperature, blending amount, and one output (y_{obs}), Syngas yield is considered (Table 3.18). Analyze the problem in the ELM framework using the Sigmoid activation function.

Solution:

Step 1: Weights connecting the input and hidden layers are randomly initialized. A set of weights characterizes each node ($\omega_1, \omega_2, \omega_3$) and a bias ($b$). For this example, one hidden layer with three nodes is considered. Three weights are randomly generated for each input feature, which are as follows (Table 3.19):

Step 2: Computation of λ (refer to Eq. 3.17):

$$\lambda = \begin{bmatrix} \Psi(\omega_1 \times x_L + b_1) & \cdots & \Psi(\omega_L \times x_1 + b_L) \\ \vdots & \cdots & \vdots \\ \Psi(\omega_1 \times x_N + b_1) & \ldots & \Psi(\omega_L \times x_N + b_L) \end{bmatrix}_{N \times L}$$

Table 3.17 MSLF computation

Dataset	x_1	x_2	y_{obs}	$\hat{y}_{pred}$	MSLF
1	2	6	4	4	0
2	3	11	7	7	0

Table 3.18 Information about input and output

Dataset	x_1	x_2	x_3	y_{obs}
1	1	2	3	5
2	2	3	4	6
3	3	4	5	7

Table 3.19 Random weight matrix

Input Node	Hidden node λ_1	Hidden node λ_2	Hidden node λ_3
x_1	0.2	0.4	0.3
x_2	0.5	0.1	0.4
x_3	0.3	0.5	0.2
b	0.1	0.2	0.3

The activation function is Sigmoid, $\Psi(z) = \frac{1}{1+e^{(-z)}}$
Similarly, by calculating all hidden layer outputs, λ is obtained.

$$\lambda = \begin{bmatrix} \psi(1\times 0.2 + 2\times 0.5 + 3\times 0.3 + 0.1) & \psi(1\times 0.4 + 2\times 0.1 + 3\times 0.5 + 0.2) & \psi(1\times 0.3 + 2\times 0.4 + 3\times 0.2 + 0.3) \\ \psi(2\times 0.2 + 3\times 0.5 + 4\times 0.3 + 0.1) & \psi(2\times 0.4 + 3\times 0.1 + 4\times 0.5 + 0.2) & \psi(2\times 0.3 + 3\times 0.4 + 5\times 0.2 + 0.3) \\ \psi(3\times 0.2 + 4\times 0.5 + 5\times 0.3 + 0.1) & \psi(3\times 0.4 + 4\times 0.1 + 5\times 0.5 + 0.2) & \psi(3\times 0.3 + 4\times 0.4 + 5\times 0.2 + 0.3) \end{bmatrix}_{N\times L}$$

$$\lambda = \begin{bmatrix} \Psi(2.2) & \Psi(2.3) & \Psi(2) \\ \Psi(3.2) & \Psi(3.3) & \Psi(3.1) \\ \Psi(4.2) & \Psi(4.3) & \Psi(3.8) \end{bmatrix}$$

$$\lambda_1 \qquad \lambda_2 \qquad \lambda_3$$

$$\lambda = \begin{bmatrix} 0.90025 & 0.90887 & 0.88080 \\ 0.96083 & 0.96443 & 0.95689 \\ 0.98523 & 0.98661 & 0.97812 \end{bmatrix}$$

Step 3: Computation of output weight vector. $\hat{\beth} = \lambda^{\dagger} y_{obs}$

where $\lambda^{\dagger} = (\lambda^T\lambda)^{-1}\lambda^T$; Here, λ^T is the transpose of λ, y_{obs} is observed output.

To solve for $\beth$, use the Moore–Penrose pseudoinverse of λ, which is denoted by $\lambda^{\dagger}$:

$$\lambda^T = \begin{bmatrix} 0.90025 & 0.96083 & 0.98523 \\ 0.90887 & 0.96443 & 0.98661 \\ 0.88080 & 0.95689 & 0.97812 \end{bmatrix}$$

$$(\lambda^T\lambda) = \begin{bmatrix} 2.70432 & 2.71690 & 2.67602 \\ 2.71690 & 2.72957 & 2.68841 \\ 2.67602 & 2.68841 & 2.64817 \end{bmatrix}$$

$$(\lambda^T\lambda)^{-1} = \frac{1}{\det(\lambda^T\lambda)}\text{Adjoint}(\lambda^T\lambda)$$

$$(\lambda^T\lambda)^{-1} = \begin{bmatrix} 210103.04337 & -147638.78660 & -62430.41641 \\ -147638.78660 & 106986.43917 & 40579.31809 \\ -62430.41641 & 40579.31809 & 21891.41874 \end{bmatrix}$$

Table 3.20 MSLF computation

Dataset	x_1	x_2	x_3	y_{obs}	$\hat{y}_{pred}$	MSLF
1	1	2	3	5	4.98977	0.000105
2	2	3	4	6	6.12951	0.016773
3	3	4	5	7	6.88268	0.013764

$$\lambda^{\dagger} = (\lambda^{T}\lambda)^{-1}\lambda^{T} = \begin{bmatrix} -27.90996 & -253.00896 & 273.47927 \\ 67.21071 & 155.09989 & -212.82836 \\ -39.69591 & 98.57442 & -59.92364 \end{bmatrix}$$

Output weight vector $\widehat{\beth} = \lambda^{\dagger} y_{obs} =$

$$\begin{bmatrix} -27.90996 & -253.00896 & 273.47927 \\ 67.21071 & 155.09989 & -212.82836 \\ -39.69591 & 98.57442 & -59.92364 \end{bmatrix} \begin{bmatrix} 5 \\ 6 \\ 7 \end{bmatrix}$$

$$= \begin{bmatrix} (-27.90996 \times 5) + (-253.00896 \times 6) + (273.47927 \times 7) \\ (67.21071 \times 5) + (155.09989 \times 6) + (-212.82836 \times 7) \\ (-39.69591 \times 5) + (98.57442 \times 6) + (-59.92364 \times 7) \end{bmatrix} = \begin{bmatrix} 256.7513 \\ -223.146 \\ -26.4985 \end{bmatrix}$$

The output is $\hat{y}_{pred} = 256.7513 \times \lambda_1 + (-223.146) \times \lambda_2 + (-26.4985) \times \lambda_3$ $\hat{y}_{pred} = 256.7513 \times \lambda_1 + (-223.146) \times \lambda_2 + (-26.4985) \times \lambda_3$.

Using the λ and output variable from the previous steps, predicting the output for the training dataset (by multiplying the $\beth$ with hidden layer output values of the input):

$$\hat{y}_1 = 256.7513 \times 0.90025 + (-223.146) \times 0.90887 + (-26.4985) \times 0.88080 = 4.98977$$
$$\hat{y}_2 = 256.7513 \times 0.96083 + (-223.146) \times 0.96443 + (-26.4985 \times 0.95689) = 6.12951$$
$$\hat{y}_3 = 256.7513 \times 0.98523 + (-223.146) \times 0.98661 + (-26.4985 \times 0.97812) = 6.88268$$

The MSLF is presented in Table 3.20.

Here, the total MSLF value is 0.010214. If not satisfied with the obtained MSLF, one option is to assign different weights. However, the procedure mentioned above remains the same.

3.7 Logistic Regression

LR uses the logistic function to predict (Pathak et al., 2020; Pradhan, 2010). The output is the chance of occurrence that a specific dataset fits into the positive or negative class (1 or 0) (Madhuri et al., 2021) (Eq. 3.18):

$$p = \frac{1}{1 + e^{-z}} \tag{3.18}$$

If the chance of occurrence is more significant than a chosen threshold, the dataset is categorized as a positive class; if it is lesser, it belongs to the negative class. Here, z is defined as (Eq. 3.19)

$$z = \omega_0 + \omega_1 x_1 + \omega_2 x_2 \ldots + \omega_n x_n \tag{3.19}$$

Here inputs are $x_1, x_2, x_3 \ldots x_n$, whereas $\omega_0, \omega_1, \omega_2, \ldots \omega_n$ are the related weights.

Optimal weights arrived during the training are expected to provide the best possible division between the classes. The negative log-likelihood is used as a loss function to decrease the cost (Eq. 3.20).

$$\text{The loss function for each chosen dataset} = -\left[y_i \log(p_i) + (1 - y_i) \log(1 - p_i)\right] \tag{3.20}$$

The subscript i in Eq. 3.20 refers to the ith training example (i = 1, 2, N). Here, the Log represents the natural logarithm. Stochastic gradient descent with a selected learning rate to minimize the loss function can be used, where one training example is considered at a time, and the weights are updated using it. Each dataset used is referred to as an iteration. Iterations can be stopped when there is no significant change in average loss per epoch in consecutive epochs. Rate of change of the loss function L with reference to ω_0, ω_1 and ω_2 can be computed (Eqs. 3.21–3.23):

$$\frac{\partial \mathrm{L}}{\partial \omega_0} = p_i - y_i \tag{3.21}$$

$$\frac{\partial \mathrm{L}}{\partial \omega_1} = x_1(p_i - y_i) \tag{3.22}$$

$$\frac{\partial \mathrm{L}}{\partial \omega_2} = x_2(p_i - y_i) \tag{3.23}$$

Now, using these values, update the values of ω_i as per Eq. 3.24:

$$\omega_i = \omega_i - \mathrm{L_r} \times \frac{\partial L}{\partial \omega_i} \tag{3.24}$$

The learning rate, $\mathrm{L_r}$, typically is chosen between 10^{-1} and 10^{-6}.

The advantages of LR are less training effort, flexibility, implementation and interpretation, effectiveness for linearly separable datasets, and less chance of over-fitting. The challenge is sensitive to noise.

Numerical problem 3.10. There are 12 datasets representing different locations in a city. Each dataset is characterized by two inputs: Distance from the Nearest Stream

(DNS) and EvapoTranspiration (ET) and one output about flooding (either flooded or non-flooded). Detailed information is presented in Table 3.21. Use LR for the analysis. Show datasets graphically for visualization. Compute flood status (flooded or non-flooded) for DNS and ET values of 6 and 8.

Solution:

The datasets are graphically visualized in Fig. 3.9.

Iteration 1: Initial value of p_i is needed to compute the rate of change of the loss function. In this regard, ω_0, ω_1 and ω_2 are all assumed as zero, making z zero as well, and are used for calculating p_i [Using Eqs. 3.18–3.19 (for DNS of 2.4 and ET of 10.2)]:

$$z = 0 + 0 \times 2.4 + 0 \times 10.2 = 0$$

$$p_i = \frac{1}{1 + e^{-z}} = \frac{1}{1 + e^{0}} = 0.5$$

Plugging in values for the first training example (DNS of 2.4, ET of 10.2, $y_i = 0$), the following values will be obtained (using Eqs. 3.21–3.23)

$$\frac{\partial \mathrm{L}}{\partial \omega_0} = p_i - y_i = 0.5 - 0 = 0.5$$

$$\frac{\partial \mathrm{L}}{\partial \omega_1} = x_1(p_i - y_i) = 2.4(0.5 - 0) = 1.2$$

Table 3.21 Datasets considered for the problem

Dataset	DNS	ET	Did flood occur?	Observed y_i
1	2.4	10.2	No	0
2	12	5.4	Yes	1
3	4.5	16	No	0
4	7.6	20.3	Yes	1
5	9.3	14.5	Yes	1
6	4.9	7.8	Yes	1
7	8.1	14.2	No	0
8	4.3	4.5	Yes	1
9	3.2	12.4	No	0
10	5.5	5.5	Yes	1
11	7.2	11.2	Yes	1
12	4.5	8.5	No	0

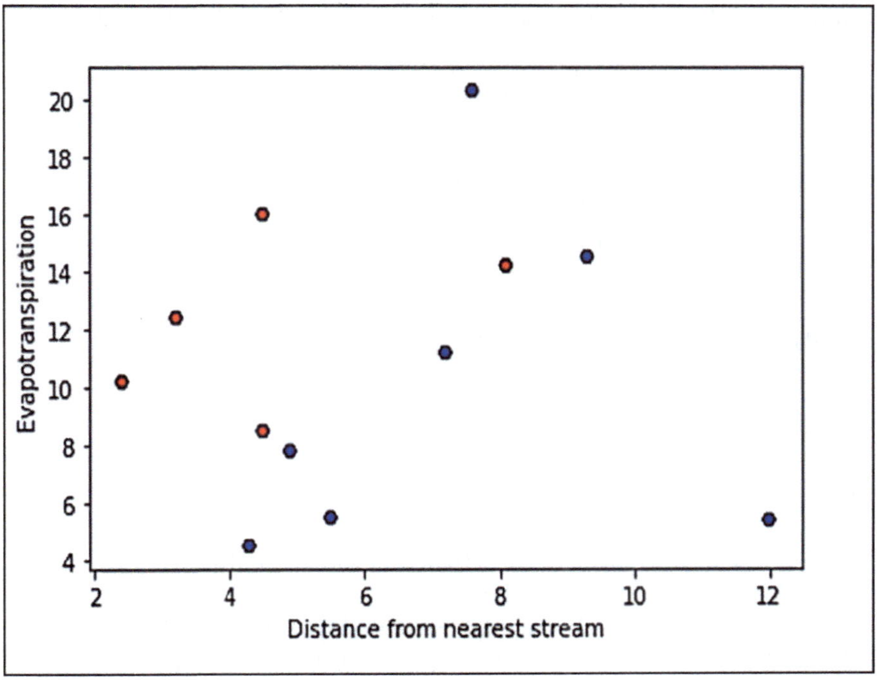

Fig. 3.9 Dataset representation for the problem (Red: Non-flooded; Blue: Flooded)

$$\frac{\partial L}{\partial \omega_2} = x_2(p_i - y_i) = 10.2(0.5 - 0) = 5.1$$

Using Eq. 3.24, update the values of ω_i which is as follows:

$$\omega_i = \omega_i - L_r \times \frac{\partial L}{\partial \omega_i}$$

Plugging in the values for ω_0,ω_1, ω_2 and assuming a L_r of 0.01,

$$\omega_0 = 0 - 0.01 \times 0.5 = -0.005$$
$$\omega_1 = 0 - 0.01 \times 1.2 = -0.012$$
$$\omega_2 = 0 - 0.01 \times 5.1 = -0.051$$

Loss for dataset 1 $= -[0 \times \log(0.5) + 1 \times \log(0.5)] = 0.6931$

Iteration 2: Calculating p_i with the second training example (DNS of 12, ET of 5.4, $y_i = 1$) using Eq. 3.18:

$$p_i = \frac{1}{1 + e^{-z}} = \frac{1}{1 + e^{-[(-0.005)+(-0.012\times 12)+(-0.051\times 5.4)]}} = 0.395$$

It is followed by calculating the gradients of the weights using Eqs. 3.21–3.23

$$\frac{\partial L}{\partial \omega_0} = (0.395 - 1) = -0.605$$

$$\frac{\partial L}{\partial \omega_1} = 12(0.395 - 1) = -7.254$$

$$\frac{\partial L}{\partial \omega_2} = 5.4(0.395 - 1) = -3.264$$

Updating the weights using Eq. 3.24:

$$\omega_0 = -0.005 - 0.01 \times -0.605 = 0.001$$

$$\omega_1 = -0.012 - 0.01 \times -7.254 = 0.061$$

$$\omega_2 = -0.051 - 0.01 \times -3.264 = -0.018$$

Loss for dataset 2 $= -[1 \times \log(0.395) + 0 \times \log(0.6045)] = 0.928$.

Similarly, all other datasets were processed, completing one epoch. Table 3.22 presents the results for the epoch 1.

The average loss is 0.919 for epoch 0. The lower the loss, the better the model has trained. A decrease in loss is observed with an increase in epochs (Table 3.23). The algorithm terminates when the difference between the average losses of two

Table 3.22 Results at epoch 1

Dataset	DNS	ET	y_i	p_i	$\frac{\partial L}{\partial \omega_0}$	$\frac{\partial L}{\partial \omega_1}$	$\frac{\partial L}{\partial \omega_2}$	ω_0	ω_1	ω_2	Loss
1	2.4	10.2	0	0.500	0.500	1.200	5.100	−0.005	−0.012	−0.051	0.693
2	12	5.4	1	0.395	−0.605	−7.254	−3.264	0.001	0.061	−0.018	0.928
3	4.5	16	0	0.495	0.495	2.227	7.919	−0.004	0.038	−0.098	0.683
4	7.6	20.3	1	0.155	−0.845	−6.419	−17.146	0.005	0.102	0.074	1.862
5	9.3	14.5	1	0.884	−0.116	−1.080	−1.684	0.006	0.113	0.091	0.123
6	4.9	7.8	1	0.781	−0.219	−1.075	−1.712	0.008	0.124	0.108	0.248
7	8.1	14.2	0	0.927	0.927	7.510	13.166	−0.001	0.049	−0.024	2.62
8	4.3	4.5	1	0.525	−0.475	−2.040	−2.135	0.003	0.069	−0.002	0.643
9	3.2	12.4	0	0.549	0.549	1.756	6.803	−0.002	0.052	−0.070	0.795
10	5.5	5.5	1	0.474	−0.526	−2.894	−2.894	0.003	0.081	−0.042	0.747
11	7.2	11.2	1	0.530	−0.470	−3.386	−5.266	0.008	0.115	0.011	0.635
12	4.5	8.5	0	0.650	0.650	2.924	5.523	0.001	0.085	−0.044	1.049
										Average loss	0.919

Table 3.23 The average loss in each epoch

Epoch	Loss
0	0.919
1	0.836
2	0.793
3	0.768
4	0.751
5	0.740
6	0.732
7	0.725
8	0.720
9	0.716
10	0.713
…	…
2915	0.630

consecutive epochs is less than or equal to 10^{-6}, which occurred at epoch 2915 (Fig. 3.10).

The values of the weights at this epoch are found to be

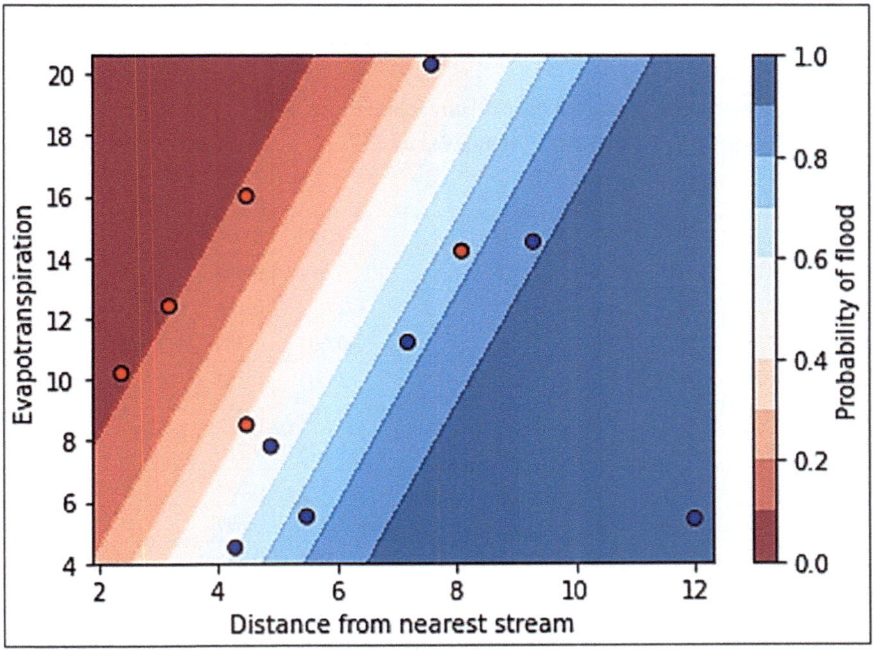

Fig. 3.10 Status of flood nodes at epoch 2915 (Red: Non-flooded; Blue: Flooded)

$$\omega_0 = -1.934, \omega_1 = 0.771, \omega_2 = -0.223$$

To predict the class (flooded or non-flooded) of an unseen dataset, the values of DNS and ET, i.e., 6 and 8, can be directly substituted in Eq. 3.19.

$$z = -1.934 + 0.771 \times 6 - 0.223 \times 8 = 0.908$$

Visualizing the classification by calculating the probability for all possible datasets (using the updated weights) as follows:

$$p_i = \frac{1}{1+e^{-z}} = \frac{1}{1+e^{-0.908}} = 0.713$$

If the output value is more significant than 0.5, the prediction is considered as flooded. Two non-flooded locations have been incorrectly classified as flooded, while one flooded location has incorrectly been classified as non-flooded (refer to Table 3.24).

Numerical problem 3.11. The dataset consists of 8 thermal power plants (Table 3.25). Inputs are TEMperature (TEM) and HUMidity (HUM), and the output is about safety (either safe or unsafe). Consider initial weights as 0.1. LR can be utilized for the analysis. Solve the problem for one epoch. Compute the status of the thermal power plant for a TEM value of 10 and a HUM value of 5.

Solution:

Table 3.24 Observed and predicted flood occurrences at epoch 2915

Dataset	DNS	ET	Flooded?	Observed y_i	z^*	Predicted probability p	Flood occurrence$^+$
1	2.4	10.2	No	0	−2.358	0.086	0
2	12	5.4	Yes	1	6.114	0.998	1
3	4.5	16	No	0	−2.033	0.116	0
4	**7.6**	**20.3**	**Yes**	**1**	−0.601	0.354	**0**
5	9.3	14.5	Yes	1	2.003	0.881	1
6	4.9	7.8	Yes	1	0.105	0.526	1
7	**8.1**	**14.2**	**No**	**0**	1.145	0.759	**1**
8	4.3	4.5	Yes	1	0.378	0.593	1
9	3.2	12.4	No	0	−2.232	0.097	0
10	5.5	5.5	Yes	1	1.080	0.746	1
11	7.2	11.2	Yes	1	1.120	0.754	1
12	4.5	8.5	No	0	−0.360	0.411	0

$^*z = -1.934 + 0.771 \times \text{DNS} - 0.223 \times \text{ET}; p = \frac{1}{1+e^{-z}}$

Table 3.25 Dataset for the problem

Dataset	TEM	HUM	Is the plant safe?	Observed y_i
1	1	30	No	0
2	38	1	Yes	1
3	1	18	No	0
4	8	13	No	0
5	46	50	No	0
6	23	16	Yes	1
7	5	26	No	0
8	47	17	No	0

Iteration 1: Initial value of p_i is needed to compute the rate of change of the loss function, i.e., when ω_0, ω_1 and ω_2 are all taken as 0.1. Calculating p_i Using Eqs. 3.19 and 3.18 (for TEM of 1 and HUM of 30)

$$z = 0.1 + 0.1 \times 1 + 0.1 \times 30 = 3.2$$

$$p_i = \frac{1}{1 + e^{-z}} = \frac{1}{1 + e^{-3.2}} = 0.9608$$

Plugging in values for the first training example (TEM of 1, HUM of 30, $y_i = 0$) and using Eqs. 3.21–3.23, the output is as follows:

$$\frac{\partial \mathrm{L}}{\partial \omega_0} = p_i - y_i = 0.9608 - 0 = 0.9608$$

$$\frac{\partial \mathrm{L}}{\partial \omega_1} = x_1(p_i - y_i) = 1(0.9608 - 0) = 0.9608$$

$$\frac{\partial \mathrm{L}}{\partial \omega_2} = x_2(p_i - y_i) = 30(0.9608 - 0) = 28.83$$

Now, using these values, the updation of values of ω_i

$$\omega_i = \omega_i - \alpha \times \frac{\partial \mathrm{L}}{\partial \omega_i}$$

Plugging in the values for ω_0,ω_1, ω_2 and assuming a learning rate α of 0.01 (using Eq. 3.24)

$$\omega_0 = 0.1 - 0.01 \times 0.9608 = 0.0904$$

$$\omega_1 = 0.1 - 0.01 \times 0.9608 = 0.0904$$

$$\omega_2 = 0.1 - 0.01 \times 28.83 = -0.1883$$

Iteration 2: Calculating p_i with the second training example (TEM of 38, HUM of 1, $y_i = 1$) using Eq. 3.18,

$$p_i = \frac{1}{1 + e^{-z}} = \frac{1}{1 + e^{-[(0.0904)+(0.0904\times 38)+(-0.1883\times 1)]}} = 0.9657$$

It is followed by calculating the gradients of the weights using Eqs. 3.21–3.23

$$\frac{\partial \mathrm{L}}{\partial \omega_0} = (0.9657 - 1) = -0.0343$$

$$\frac{\partial \mathrm{L}}{\partial \omega_1} = 38(0.9657 - 1) = -1.3043$$

$$\frac{\partial \mathrm{L}}{\partial \omega_2} = 1(0.9657 - 1) = -0.0343$$

Updating the weights using Eq. 3.24:

$$\omega_0 = 0.0904 - 0.01 \times -0.0343 = 0.0907$$

$$\omega_1 = 0.0904 - 0.01 \times -1.3043 = 0.1034$$

$$\omega_2 = -0.1883 - 0.01 \times -0.0343 = -0.1879$$

Similarly, all other datasets were processed, completing one epoch. Table 3.26 presents the results for the epoch 1.

Table 3.26 Results at epoch 1

Dataset	TEM	HUM	y_i	p_i	$\frac{\partial L}{\partial \omega_0}$	$\frac{\partial L}{\partial \omega_1}$	$\frac{\partial L}{\partial \omega_2}$	ω_0	ω_1	ω_2
1	1	30	0	0.961	0.9608	0.9608	28.8250	0.0904	0.0904	−0.1883
2	38	1	1	0.966	−0.0343	−1.3043	−0.0343	0.0907	0.1034	−0.1879
3	1	18	0	0.040	0.0396	0.0396	0.7130	0.0903	0.1030	−0.1950
4	8	13	0	0.165	0.1651	1.3207	2.1462	0.0887	0.0898	−0.2165
5	46	50	0	0.001	0.0014	0.0622	0.0677	0.0887	0.0892	−0.2172
6	23	16	1	0.208	−0.7916	−18.2058	−12.6649	0.0966	0.2713	−0.0905
7	5	26	0	0.289	0.2889	1.4445	7.5112	0.0937	0.2568	−0.1656
8	47	17	0	1.000	0.9999	46.9959	16.9985	0.0837	−0.2131	−0.3356

$$z = 0.0837 + (-0.2131 \times \text{TEM}) + (-0.3356 \times \text{HUM})$$
$$= 0.0837 + (-0.2131 \times 10) + (-0.3356 \times 5) = -3.7253$$

$p_i = \frac{1}{1+e^{-z}} = \frac{1}{1+e^{3.7253}} = 0.0235$ indicating unsafe plant (based on the threshold of 0.5).

3.8 K-Nearest Neighbours

KNN stores all the datasets and classifies new ones built on distance functions (Alfeilat et al., 2019; Modaresi et al., 2018; Uddin et al., 2022). Distance from the testing dataset to each training dataset can be computed using Eq. 3.25 (Madhuri et al., 2021):

$$\text{D}(\text{X}, \text{X}_i) = \left(\sum_{j=1}^{n} \left|x_j - x_{ij}\right|^p\right)^{1/p} \quad (3.25)$$

where X $(x_1, x_2, x_3 \ldots x_n)$ are testing datasets, $\text{X}_i(x_{i1}, x_{i2}, x_{i3} \ldots x_{in})$ are training datasets ($i = 1$ to N), n is the number of features, p is the Minkowski metric, and N is the number of training examples. It is aimed to predict the class of the testing dataset X using the training datasets. KNN does not require training. The entire dataset is presented to the algorithm, and new predictions are made based on distance measures. Then, the most suitable class is selected for the training dataset among K-nearest neighbours to the testing dataset.

The advantages are fewer parameter requirements and no impact of adding or removing datasets.

Numerical problem 3.12. The dataset presented in Table 3.27 is related to flood occurrence and is characterized by two input variables, DNS and ET. Employ KNN technique. Identify the closest K-nearest neighbours for DNS of 9 and ET of 13.

Solution:

KNN operates on an elementary principle. The closest K-neighbours decide the class of a new dataset. The newer dataset (9-unit DNS and 13-unit ET) will be classified as flooded since the three closest datasets to it are of the positive class when K = 3 (Fig. 3.11). A similar computation can be done for K = 4 or K = 5. However, one or more values of K may provide optimal results.

To verify the above claim, the Euclidean distance ($p = 2$ in Eq. 3.25) between training and the testing dataset can be found, and the distances in ascending order (Table 3.28).

Distance of (8.1, 14.2) from (9, 13)

$$\sqrt{(8.1-9)^2 + (14.2-13)^2} = 1.5$$

Table 3.27 Datasets related to flood occurrence

Dataset	DNS	ET	Did flood occur?	Observed y_i
1	2.4	10.2	No	0
2	12	5.4	Yes	1
3	4.5	16	No	0
4	7.6	20.3	Yes	1
5	9.3	14.5	Yes	1
6	4.9	7.8	Yes	1
7	8.1	14.2	No	0
8	4.3	4.5	Yes	1
9	3.2	12.4	No	0
10	5.5	5.5	Yes	1
11	7.2	11.2	Yes	1
12	4.5	8.5	No	0

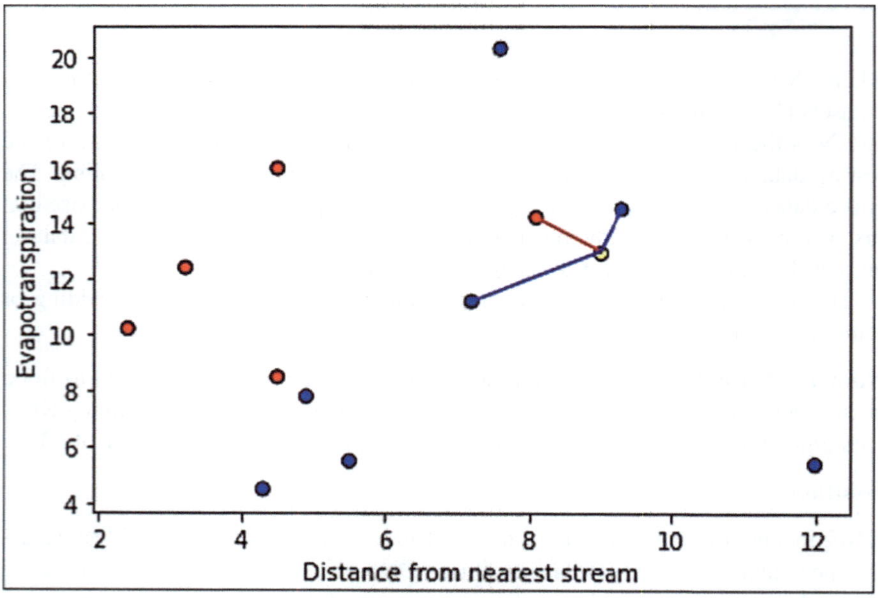

Fig. 3.11 KNN-three nearest neighbours (Red: Non-flooded; Blue: Flooded)

As evident from Table 3.28, amongst the first three closest training datasets, 2 are of the 1 class, i.e., flooded, and thus, the testing dataset is also assigned the class of 1.

Numerical problem 3.13. The dataset is related to Solar power plants linked to renewal energy engineering (Table 3.29). Temperature (TEM) and humidity (HUM)

Table 3.28 Distance of test dataset from each training dataset

DNS	ET	y_i	Distance
8.100	14.2	0	1.50
9.300	14.5	1	1.53
7.200	11.2	1	2.55
4.500	16	0	5.41
3.200	12.4	0	5.83
4.5	8.5	0	6.36
4.9	7.8	1	6.62
2.4	10.2	0	7.17
7.6	20.3	1	7.43
12	5.4	1	8.17
5.5	5.5	1	8.28
4.3	4.5	1	9.71

are the inputs, whereas the status of the plant is the output. Employ KNN technique. Identify the closest K-nearest neighbours for TEM of 44 and HUM of 46.

Solution:

The closest K-neighbours decide the class of a new dataset. In this case, the new dataset is 44-unit TEM and 46-unit HUM. To verify the above claim, the Euclidean distance with $p = 2$ (Eq. 3.25) from the testing dataset (or new dataset) to every

Table 3.29 Dataset related to Solar power plants

Dataset	TEM	HUM	Is the plant is safe?	Observed y_i
1	1	30	No	0
2	38	1	Yes	1
3	1	18	No	0
4	8	13	No	0
5	46	50	No	0
6	23	16	Yes	1
7	5	26	No	0
8	47	17	No	0
9	43	29	Yes	1
10	37	2	Yes	1
11	21	37	Yes	1
12	45	43	No	0
13	50	23	Yes	1
14	44	25	Yes	1
15	10	36	Yes	1

Table 3.30 Distance of test dataset from each training dataset

TEM	HUM	y_i	Distance
45	43	0	3.16
46	50	0	4.47
43	29	1	17.03
44	25	1	21.00
50	23	1	23.77
21	37	1	24.70
47	17	0	29.15
10	36	1	35.44
23	16	1	36.62
5	26	0	43.83
37	2	1	44.55
38	1	1	45.40
1	30	0	45.88
8	13	0	48.84
1	18	0	51.31

training dataset can be estimated. Arrange the distances in ascending order (column 4: Table 3.30).

Distance of (45, 43) from (44, 46): $\sqrt{(45-44)^2+(43-46)^2} = 3.16$

As evident from column 4 of Table 3.30, the first two closest training datasets, 2 are of the 0 class, i.e., unsafe, and thus, the testing dataset is also assigned the class of 0.

Revision Questions and Exercise Problems

3.1. One input layer (humidity, temperature, rainfall, wind speed, and road surface temperature) and one hidden layer with one node exist. Input values are 0.18, 0.36, 0.54, 0.68, and 0.82. Assume weights between inputs and hidden node as 0.34, 0.54, 0.64, 0.74, and 0.84. Compute output from the hidden layer using the ReLU and hyperbolic tangent activation functions. Comment on the output.

3.2. What is FFBP-ANN? Explain the philosophy behind the same.

3.3. Five inputs, A to E, with values of 3, 4, 6, 8, and 9, produce an output of 0.8. It is suggested that one hidden layer be included with one node. Initial weights between inputs and hidden are 0.2, 0.45, 0.65, and 0.85, whereas it is 0.4 in the case of hidden to output. Establish a relationship using FFBP-ANN with Sigmoid as the activation function. Show the computations for one epoch with a learning rate of 0.24.

3.4. What is WNN? Explain the philosophy behind the same.

3.5. What is the decomposition principle employed in WNN?

3.6. What are the parameters that govern the WNN?

3.7. What are mother wavelets and their functions? Mention the names of three mother wavelets.

3.8. Five datasets with two input variables (x_1 & x_2) and one output (y) are related to biomedical engineering (Table 3.31). Relate inputs and output utilizing WNN.

3.9. What is SVR? Explain the philosophy behind the same.

3.10. What are the parameters that govern SVR?

3.11. What is the Kernel function? What types of kernels can be employed while working on SVR?

3.12. What is a hyperplane in SVR?

3.13. What are the advantages and disadvantages of SVR?

3.14. Table 3.32 presents five datasets, with input (x) and output (y) in the highway alignment framework. Establish a relationship between them using SVR.

3.15. What is ELM? Explain the philosophy behind the same.

3.16. What are the advantages and disadvantages of ELM?

3.17. Three datasets were developed experimentally, with three input variables (x_1, x_2, x_3) and one output variable (y_{obs}). Analyze the problem in the ELM framework (refer to Table 3.33):

Table 3.31 Information about datasets

Dataset	x_1	x_2	y
1	2	1	4
2	3	4	8
3	4	7	12
4	8	9	16
5	9	11	22

Table 3.32 Information about datasets

Dataset	x	y
1	1.2	4.2
2	2.1	7.1
3	3.3	5.3
4	4.3	9.3
5	5.3	10.3

Table 3.33 Information about datasets

Dataset	x_1	x_2	x_3	y_{obs}
1	1.2	2.2	3.3	5.5
2	2.1	3.1	4.1	6.1
3	3.2	4.2	5.3	7.3

3.18. What is the physical significance of weights in LR?
3.19. What is the function of non-linear transformation in LR?
3.20. What is meant by epoch and iteration in the context of LR?
3.21. Mention one distinct advantage in LR that affects the accuracy of the outcome.
3.22. What is KNN? Explain the philosophy behind the same.
3.23. What is the physical meaning of K in KNN?
3.24. Does KNN require any training? Justify your response logically.
3.25. Mention one distinct advantage in KNN that affects the accuracy of the outcome.
3.26. Solve the problem utilizing LR and KNN. Data is presented in Table 3.34. You can assume suitable data, if any.

Advanced Review Questions

3.27. Why is the activation function also termed as a transfer function?
3.28. Can you propose a new activation function and justify its utility over the existing one?
3.29. What is over-fitting and under-fitting? Can you minimize the same?
3.30. What are discrete and continuous mother wavelets? Explain their suitability.
3.31. Why is LR preferred over linear regression for binary classification problems?
3.32. Is normalization of the dataset necessary?
3.33. What is the difference between feature scaling and normalization?
3.34. What is the concept of the lazy learner in KNN?
3.35. What is the basis for optimal K in the KNN? What is the implication of choosing the small values of K?

Table 3.34 Information about features and flooding status

Dataset	DNS	ET	y_i	Flooded?
1	2.8	10.8	1	Yes
2	12.6	5.8	0	No
3	4.8	16.8	1	Yes
4	7.8	20.8	0	No
5	9.9	14.8	1	Yes
6	6.9	8.8	0	No
7	8.8	16.2	0	No
8	4.8	4.8	1	Yes
9	3.8	12.8	1	Yes
10	6.5	5.8	1	Yes
11	7.8	11.8	0	No
12	4.8	8.8	0	No

References

Alfeilat, H. A. A., Hassanat, A. B. A., Lasassmeh, O., Tarawneh, A. S., Alhasanat, M. B., Eyal Salman, H. S., & Prasath, V. B. S. (2019). Effects of distance measure choice on K-nearest neighbor classifier performance: A review. *Big Data, 7*, 221–248.

Granata, F., Gargano, R., & De Marinis, G. (2016). Support vector regression for rainfall-runoff modeling in urban drainage: A comparison with the EPA's storm water management model. *Water, 8*, 69.

Guo, T., Zhang, T., Lim, E., López-Benítez, M., Ma, F., & Yu, L. (2022). A review of wavelet analysis and its applications: Challenges and opportunities. *IEEE Access, 10*, 58869–58903.

Huang, G., Huang, G. B., Song, S., & You, K. (2015). Trends in extreme learning machines: A review. *Neural Networks, 61*, 32–48.

Madhuri, R. (2022). *Risk assessment and mitigation strategies for urban floods under climate change.* Ph.D. thesis, BITS Pilani, India.

Madhuri, R., Sistla, S., & Raju, K. S. (2021). Application of machine learning algorithms for flood susceptibility assessment and risk management. *Journal of Water and Climate Change, 12*, 2608–2623.

Modaresi, F., Araghinejad, S., & Ebrahimi, K. (2018). A comparative assessment of artificial neural network, generalized regression neural network, least-square support vector regression, and K-nearest neighbor regression for monthly streamflow forecasting in linear and non-linear conditions. *Water Resources Management, 32*, 243–258.

Mohammadi, B. (2021). A review on the applications of machine learning for runoff modeling. *Sustainable Water Resources Management, 7*, 98.

Pathak, A. K., Sharma, M., Katiyar, S. K., Katiyar, S., & Nagar, P. K. (2020). Logistic regression analysis of environmental and other variables and incidences of tuberculosis in respiratory patients. *Scientific Reports, 10*, 21843.

Pradhan, B. (2010). Flood susceptible mapping and risk area delineation using logistic regression, GIS and remote sensing. *Journal of Spatial Hydrology, 9*, 1–18.

Raghavendra, N. S., & Deka, P. C. (2014). Support vector machine applications in the field of hydrology: A review. *Applied Soft Computing, 19*, 372–386.

Rao, G. S. (2000). Artificial neural networks in hydrology. i: Preliminary concepts. *Journal of Hydrologic Engineering, 5*, 115–123.

Sharma, S. (2017). Activation functions in neural networks. Accessed February 10, 2024, https://towardsdatascience.com/activation-functions-neural-networks-1cbd9f8d91d6
Sujay, R. N., & Paresh, C. D. (2014). Support vector machine applications in the field of hydrology: A review. *Applied Soft Computing, 19*, 372–386.
Uddin, S., Haque, I., Lu, H., Moni, M. A., & Gide, E. (2022). Comparative performance analysis of K-nearest neighbour (KNN) algorithm and its different variants for disease prediction. *Scientific Reports, 12*, 6256.
Vapnik, N. (1998). *Statistical learning theory*. New York: John Wiley & Sons.
Vogeti, R. K., Mishra, B. R., & Raju, K. S. (2022). Machine learning algorithms for streamflow forecasting of lower Godavari basin. *H2Open Journal*, 5, 670–685.
Vogl, M., Rötzel, P. G., & Homes, S. (2022). Forecasting performance of wavelet neural networks and other neural network topologies: A comparative study based on financial market data sets. *Machine Learning with Applications, 8*, 100302.
Wang, J., Lu, S., Wang, S. H., & Zhang, Y. D. (2022). A review on extreme learning machine. *Multimedia Tools and Applications, 81*, 41611–41660.
Yang, X., Kumehara, H., & Zhang, W. (2009). Back propagation wavelet neural network-based prediction of drill wear from thrust force and cutting torque signals. *Computer and Information Science, 2*, 75–86.

Suggested Further Reading

Distance Measures for Machine Learning. Accessed July 18, 2021, https://machinelearningmastery.com/distance-measures-for-machine-learning/
Example KNN: The nearest neighbor algorithm. Accessed 18 July 2021, https://www.scss.tcd.ie/~koidlk/cs4062/ExampleKNN.pdf
Feng, B., Xu, Y., Zhang, T., & Zhang, X. (2022). Hydrological time series prediction by extreme learning machine and sparrow search algorithm. *Water Supply, 22*, 3143–3157.
Kshirsagar, M. P., & Khare, K. C. (2023). Support vector regression models of storm water quality for a mixed urban land use. *Hydrology, 10*, 66.
Yaseen, Z. M., Sulaiman, S. O., Deo, R. C., & Chau, K. W. (2019). An enhanced extreme learning machine model for river flow forecasting: State-of-the-art, practical applications in water resource engineering area and future research direction. *Journal of Hydrology, 569*, 387–408.

Chapter 4
Advanced Machine Learning Algorithms

4.1 Introduction

The chapter describes a few advanced ML algorithms specifically employed for forecasting. They are Convolutional Neural Networks (CNN), Recurrent Neural Networks (RNN), Long Short-Term Memory (LSTM), Bi-directional (Bi)-LSTM, Gated Recurrent Unit (GRU), and possible hybridizations of these algorithms. The chapter also discusses Boosting Algorithms, viz., Adaptive Boosting (AdaBoost), eXtreme Gradient Boosting (XGBoost), and Categorical Boosting (CatBoost).

4.2 Convolutional Neural Networks

CNN is established on the philosophy of local neural connectivity motivated by the cognitive process of the animal visual cortex (Islam et al., 2022; Neu et al., 2022). Figure 4.1 presents the architecture of CNN (Van et al., 2020). Details of CNN are presented in Vogeti et al. (2024).

- Layer-I: The input layer feeds the data to the model. The space-specific and time-specific details from the previous are utilized to correlate the observed variable. Information about input variables at distinct times is the basis for components of the matrix.
- Layer-II: The convolution layer which affects significantly feature extraction, consisting of neurons (*features refer to the dimensionality of output feature maps after applying convolution and pooling operations; neurons per layer determine the capacity of the model to capture and represent data, the range is 1–256*). It has a convoluted matrix, a product of filter (or kernel), and input matrices. Filters in a convolutional layer determine the number of unique features the layer can detect, such as edges, textures, or complex patterns in the input data. Balancing

K. Srinivasa Raju and D. Nagesh Kumar, *Artificial Intelligence and Machine Learning Techniques in Engineering and Management*,
https://doi.org/10.1007/978-981-96-2621-2_4

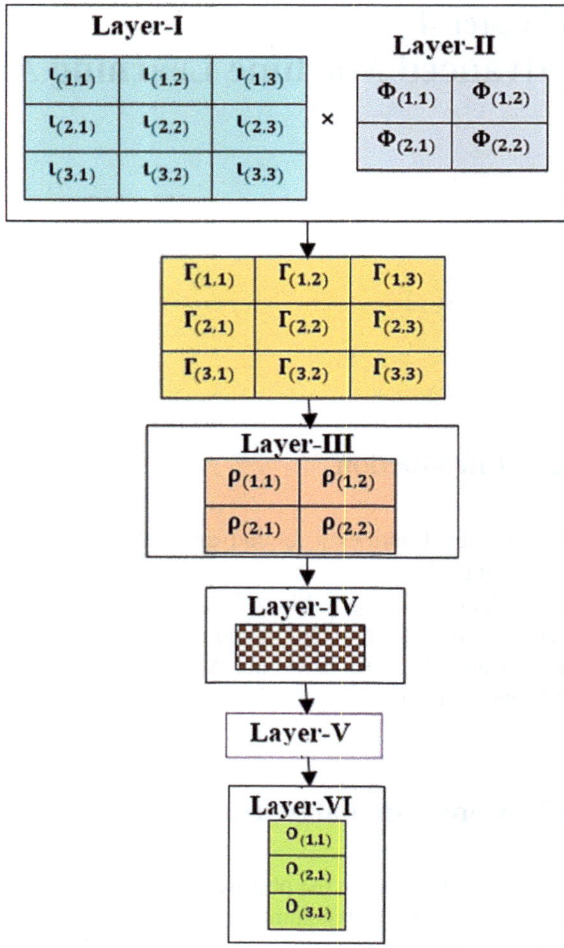

Fig. 4.1 Architecture of CNN

the number of kernels ensures sufficient feature detection capability. It is ensured by starting with fewer kernels in the initial layers and increasing them gradually in deeper layers. Related expression for *i*th data is expressed as (Eq. 4.1)

$$\Gamma_i(l, m) = \sum_{a=1}^{N_{\Phi H}} \sum_{b=1}^{N_{\Phi W}} \left(\iota_{l+a-1,\, m+b-1}(i) \times \Phi_{a,\, b}(i)\right) + \mathrm{Bias} \tag{4.1}$$

where $\Gamma_i(l, m)$ is the convolutional layer for ith data; l and m refer to the row and column features of the convoluted matrix; $N_{\Phi H}$ and $N_{\Phi W}$, respectively, the height and width of the filter; $\Phi_{a,b}(i)$ represents the filter matrix element of ath row and bth column for ith data; $\times$ denotes the dot product; and $\iota_{l+a-1,\, m+b-1}(i)$ represents the input matrix element of $(l + a - 1)$th row and $(m + b - 1)$th column for ith data.

- Layer-III: The pooling layer is a down-sampling operation that reduces the dimensionality of feature maps, thus decreasing computational complexity and minimizing over-fitting. Max pooling (selects the maximum value in each pooling window, preserving prominent features, and introducing spatial invariance) and average pooling (computes the average value, retaining more contextual information) are the frequently utilized types of pooling operations. This operation reduces the number of learnable parameters by estimating the maximum or average value of each row and captures the most pertinent features in the input layer (Eq. 4.2):

$$A\rho_i = \text{avg}\Gamma_i(l, m), \text{ where } l \in N_{\Phi H} \text{ and } m \in N_{\Phi W} \tag{4.2}$$

where avg $\Gamma_i(l, m)$ represents the average pooled layer for ith data.

- Layer-IV: In a flatten layer, the pooled features are metamorphosed into a one-dimensional vector and sent to the fully connected layer.
- Layer-V: In a fully connected layer, different features that were learned by the convolutional, pooling and flatten layers are metamorphosed into a dense vector, and their corresponding elements in this layer are expressed as (Eqs. 4.3–4.5):

$$\zeta_i^1 = \omega_{11} A\rho_1 + \omega_{01} \tag{4.3}$$

$$\zeta_i^2 = \omega_{12} A\rho_2 + \omega_{02} \tag{4.4}$$

$$\zeta_i^3 = \omega_{13} A\rho_3 + \omega_{03} \tag{4.5}$$

where ζ_i^1, ζ_i^2, and ζ_i^3 represent the first, second, and third elements of the dense one-dimensional vector of the fully connected layer; ω_{11}, ω_{12}, ω_{13} are the weights associated with the elements of dense vector; and ω_{01}, ω_{02}, ω_{03} are the biases associated with the elements of dense vector.

- Layer-VI: The last is the output layer (Eq. 4.6):

$$y_{p,i}^c = \frac{e^{\zeta_i^c}}{\sum_{c'=1}^{N_{SM}} e^{\zeta_i^{c'}}} \tag{4.6}$$

where $y_{p,i}^c$ is the output element estimated for ith data; N_{SM} is the number of Softmax units (if the Softmax function is employed); and ζ_i^c and $\zeta_i^{c'}$ are components of a dense, fully connected layer, and the loss function is computed. The process continues until the termination criterion is achieved (Alzubaidi et al., 2021). Figure 4.2 presents the workflow of CNN.

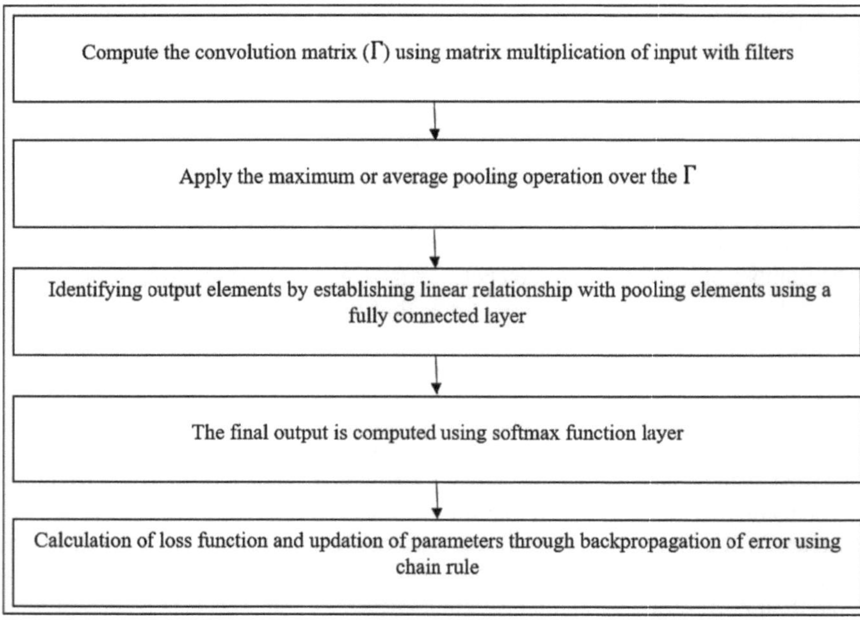

Fig. 4.2 Workflow of CNN

Other parameters that govern the training process are batch size, activation function, learning rate, epochs, and dropout. The challenge in CNN is its rigid structure for accommodating an adequate number of parameters, which are susceptible to over-fitting.

Numerical Problem 4.1. Apply the CNN to relate input (prostate volume, age) and output (risk factors for prostate cancer). Use the average pooling method. Refer to Fig. 4.2 to understand the working steps of the problem.

	Input matrix		Observed output matrix
	x_1	x_2	O
Dataset 1	4	2	0.5
Dataset 2	6	4	0.75

Solution:

Step 1: Compute Γ using matrix multiplication of input with filters.

The Γ could be computed by matrix multiplication of the input matrix (ι) with the appropriate filter matrix (Φ) that is randomly generated.

Γ		=	ι		×	Φ	
Γ (1,1)	Γ (1,2)		ι (1,1)	ι (1,2)		Φ (1,1)	Φ (1,2)
Γ (2,1)	Γ (2,2)		ι (2,1)	ι (2,2)		Φ (2,1)	Φ (2,2)

The convolutional layer of ith data is mathematically expressed as

$$\Gamma_i(l, m) = \sum_{a=1}^{N_{\Phi H}} \sum_{b=1}^{N_{\Phi W}} \left(\iota_{l+a-1,\ m+b-1}(i) \times \Phi_{a,b}(i)\right)$$

The size of the $\Gamma_i(l, m)$ is denoted by $(N_{\Phi W},\ N_{\Phi H})$ and is computed as $N_{\Phi W} = N_{\iota W} - N_{\Phi W} + 1$ and $N_{\Phi H} = N_{\iota H} - N_{\Phi H} + 1$.

For the given problem, the randomly generated Φ is multiplied by the ι as follows:

ι		×	Φ	
4	2		0.4	0.6
6	4		0.6	0.4

In this case $N_{\iota H} = N_{\iota W} = N_{\Phi H} = N_{\Phi W} = 2$.

The output matrix should have a size of 2. In this case, the size of convolution is reduced to 1; thus, it is padded with zero elements. Thus, $N_{PW} = 1$ and $N_{PH} = 1$ are the width and height of the zero padding elements, which are added to find the size of the Γ and are computed as

$$\begin{aligned} N_{\Gamma W} &= N_{\iota W} - N_{\Phi W} + N_{PW} + 1 \\ &= 2 - 2 + 1 + 1 = 2 \end{aligned}$$

$$\begin{aligned} N_{\Gamma H} &= N_{\iota H} - N_{\Phi H} + N_{PH} + 1 \\ &= 2 - 2 + 1 + 1 = 2. \end{aligned}$$

The convolutional matrix size is $(N_{\Gamma W}, N_{\Gamma H}) = (2,2)$

$$\Gamma(1, 1) = 4(0.4) + 2(0.6) = 1.6 + 1.2 = 2.8$$

$$\Gamma(1, 2) = 4(0.6) + 2(0.4) = 2.4 + 0.8 = 3.2$$

$$\Gamma(2, 1) = 6(0.4) + 4(0.6) = 2.4 + 2.4 = 4.8$$

$$\Gamma(2, 2) = 6(0.6) + 4(0.4) = 3.6 + 1.6 = 5.2$$

$\Gamma =$	2.8	3.2
	4.8	5.2

Step 2 Apply the average pooling (Aρ) method over the Γ:

$$A\rho = \frac{2.8 + 3.2 + 4.8 + 5.2}{4} = 4.$$

Step 3 Identifying two output variables by initializing the weights randomly through a fully connected layer:

The weights ω_{01}, ω_{11}, ω_{02}, and ω_{12} are randomly initialized to obtain the output.

Consider $\omega_{01} = \omega_{11} = 0.5$

Consider $\omega_{02} = \omega_{22} = 0.25$

$$\zeta_i^1 = \omega_{11} A\rho + \omega_{01} = 0.5(4) + 0.5 = 2.5$$

$$\zeta_i^2 = \omega_{12} A\rho + \omega_{02} = 0.25(4) + 0.25 = 1.25.$$

Step 4 Final output computed utilizing the Softmax function:

The output elements obtained are transmitted through the Softmax layer, which produces two units and is expressed as

$$y_{p,i}^c = \frac{e^{\zeta_i^{c1}}}{\sum_1^n e^{\zeta_i^{cn}}}$$

In this case, the output elements computed are

$$y_{p,1}^c = \frac{e^{2.5}}{e^{2.5} + e^{1.25}} = \frac{12.182}{12.182 + 3.4903} = 0.7772$$

$$y_{p,2}^c = \frac{e^{1.25}}{e^{2.5} + e^{1.25}} = \frac{3.4903}{12.182 + 3.4903} = 0.2228.$$

Step 5 Calculation of loss function and updation of parameters through back-propagation of error using the chain rule:

$$E_1 = O_1 - y_{p,1}^c = 0.5 - 0.7772 = -0.2772$$

$$E_2 = O_2 - y_{p,2}^c = 0.75 - 0.2228 = 0.5272$$

Minimization of the loss function is the prime objective. The parameters are updated with the back-propagation algorithm until the termination criterion is achieved.

Numerical Problem 4.2. Apply the CNN to establish a relationship between confining pressure, tensile strength, and strength of intact rocks. Use the maximum pooling method. Refer to Fig. 4.2 to understand the working steps of the problem.

Input matrix		Observed output matrix
8	5	0.4
12	8	0.2

Solution:

Step 1: Compute Γ using matrix multiplication of input with filters.

Γ		=	ι		×	Φ	
Γ (1,1)	Γ (1,2)		ι (1,1)	ι (1,2)		Φ (1,1)	Φ (1,2)
Γ (2,1)	Γ (2,2)		ι (2,1)	ι (2,2)		Φ (2,1)	Φ (2,2)

The convolutional layer of ith data is mathematically expressed as

$$\Gamma_i(l, m) = \sum_{a=1}^{N_{\Phi H}} \sum_{b=1}^{N_{\Phi W}} \left(\iota_{l+a-1,\, m+b-1}(i) \times \Phi_{a,b}(i)\right)$$

The size of the $\Gamma_i(l, m)$ is denoted by $(N_{\Phi W}, N_{\Phi H})$ and are computed as $N_{\Phi W} = N_{\iota W} - N_{\Phi W} + 1$ and $N_{\Phi H} = N_{\iota H} - N_{\Phi H} + 1$.

For the given problem, the randomly generated Φ is multiplied by the ι as follows:

ι		×	Φ	
8	5		0.55	0.45
12	8		0.45	0.55

In this case $N_{\iota H} = N_{\iota W} = N_{\Phi H} = N_{\Phi W} = 2$.

The output matrix should have a size of 2. In this case, the size of convolution is reduced to 1; thus, it is padded with zero elements. Thus, $N_{PW} = 1$ and $N_{PH} = 1$ are the width and height of the zero padding elements added to find the size of the Γ and is computed as

$$\begin{aligned} N_{\Gamma W} &= N_{\iota W} - N_{\Phi W} + N_{PW} + 1 \\ &= 2 - 2 + 1 + 1 = 2 \end{aligned}$$

$$\begin{aligned} N_{\Gamma H} &= N_{\iota H} - N_{\Phi H} + N_{PH} + 1 \\ &= 2-2+1+1 = 2. \end{aligned}$$

The Γ is $(N_{\Gamma W}, N_{\Gamma H}) = (2,2)$

$$\Gamma(1,1) = 8(0.55) + 5(0.45) = 4.4 + 2.25 = 6.65$$

$$\Gamma(1,2) = 8(0.45) + 5(0.55) = 3.6 + 2.75 = 6.35$$

$$\Gamma(2,1) = 12(0.55) + 8(0.45) = 6.6 + 3.6 = 10.2$$

$$\Gamma(2,2) = 12(0.45) + 8(0.55) = 5.4 + 4.4 = 9.8$$

Γ =	6.65	6.35
	10.2	9.8

Step 2 Apply the maximum pooling (Mρ) method over the Γ:

$M\rho = 10.2.$

Step 3 Identifying two output variables by initializing the weights through a fully connected layer:

The weights ω_{01}, ω_{11}, ω_{02}, and ω_{12} are randomly initialized to obtain output elements.

Consider $\omega_{01} = \omega_{11} = 0.4$
Consider $\omega_{02} = \omega_{12} = 0.3$

$$\zeta_i^1 = \omega_{11} M\rho + \omega_{01} = 0.4(10.2) + 0.4 = 4.48$$

$$\zeta_i^2 = \omega_{12} M\rho + \omega_{02} = 0.3(10.2) + 0.3 = 3.36.$$

Step 4 The final output is computed using the Softmax function:

The output elements obtained are transmitted through the Softmax layer, which produces two units and is expressed as

$$y_{p,i}^c = \frac{e^{\zeta_i^{c_1}}}{\sum_1^n e^{\zeta_i^{c_n}}}$$

In this case, the output elements computed are

$$y^c_{p,1} = \frac{e^{4.48}}{e^{4.48} + e^{3.36}} = \frac{88.2347}{88.2347 + 28.7892} = 0.754$$

$$y^c_{p,2} = \frac{e^{3.36}}{e^{4.48} + e^{3.36}} = \frac{28.7892}{88.2347 + 28.7892} = 0.246.$$

Step 5 Calculation of loss function and updation of parameters through back-propagation of error using the chain rule:

$$E_1 = y^c_{p,1} - O_1 = 0.754 - 0.4 = 0.354$$

$$E_2 = y^c_{p,2} - O_2 = 0.246 - 0.2 = 0.046.$$

4.3 Recurrent Neural Networks

RNN (refer to Fig. 4.3) is a variant of ANN specifically developed to examine time series data (Orojo et al., 2023). It possesses connections that facilitate the propagation of information in the form of data from one time step to the subsequent one, in contrast to conventional feed-forward neural networks. During each iteration, the algorithm receives an input that modifies its internal state, which signifies the memory of the network. This process enables the extraction of dependencies and trends in the sequential data. It generates output based on preceding hidden state information.

Consider x_t, λ_t, and y_t are input, hidden, and output states at time step t. Computations of λ_t and y_t are presented in Eqs. 4.7–4.8.

$$\lambda_t = \sigma(\omega_\lambda \times \lambda_{t-1} + \omega_x \times x_t + b_\lambda) \quad (4.7)$$

$$y_t = \sigma\left(\omega_y \times \lambda_t + b_y\right) \quad (4.8)$$

where, ω_λ, ω_x, and ω_y represent the weight matrices of hidden, input, and output states; λ_{t-1} indicates previously hidden state information; b_λ and b_y represent the bias for the hidden and output state; and σ represents the activation function. During training, the RNN learns weight matrices (ω_x, ω_λ, ω_y) and the bias terms (b_λ, b_y) by optimizing a specific loss function. One challenge with this algorithm is the vanishing gradient, which diminishes over time. In this situation, it is challenging for the network to seize long-term dependencies.

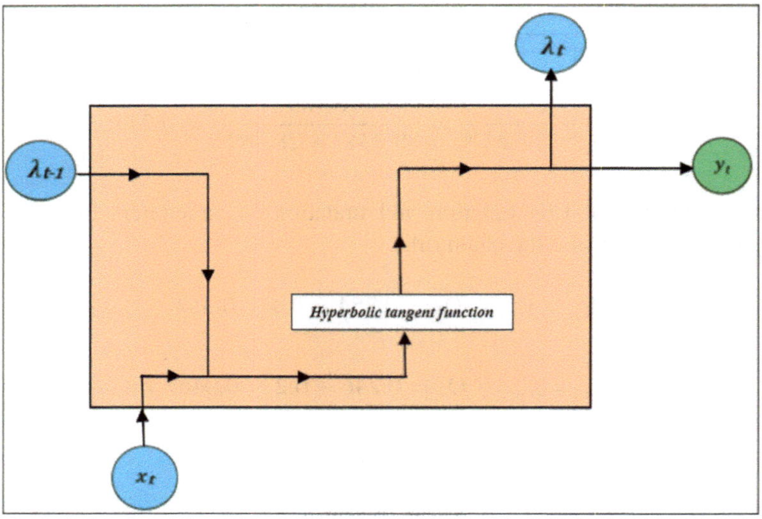

Fig. 4.3 Architecture of RNN

4.4 Long Short-Term Memory

LSTM is a sub-category of RNN that utilizes memory blocks. These act as nodes in the hidden layers and are traversed through gating units (Yu et al., 2019). It demonstrates a more remarkable ability to seize long-term dependencies from a complex time series (Horchreiter & Schmidhuber, 1997). It can overcome the problem of oscillating weights and enormous computational time realized due to vanishing and exploding gradients (Vogeti et al., 2024). LSTM architecture is presented in Fig. 4.4 (Van Houdt et al., 2020).

i. Firstly, the network receives input information of x_{t-1}, and x_t at previous and current time stamps t–1 and t, respectively. λ_{t-1} is the hidden state information retrieved from (t–1). The forget gate decides the amount of information discarded from t–1 and t, respectively. This information retention is done by combining the product of α_{fg} and λ_{t-1}, product of ω_{fg} and x_t, and a bias b_{fg}. A bias is added to enhance the flexibility and training stability in handling the information flow in the network. Application of activation function (σ) produces values between 0 and 1. A value close to 1 (or 0) means most information from the previous state is retained (or discarded). For example, a value of 0.4 indicates retention of 40% of input information in the network. Information passing through the forget gate (Eq. 4.9):

$$\text{Forget gate}\left(fg_t\right) = \sigma\left(\omega_{fg}x_t + \alpha_{fg}\lambda_{t-1} + b_{fg}\right). \tag{4.9}$$

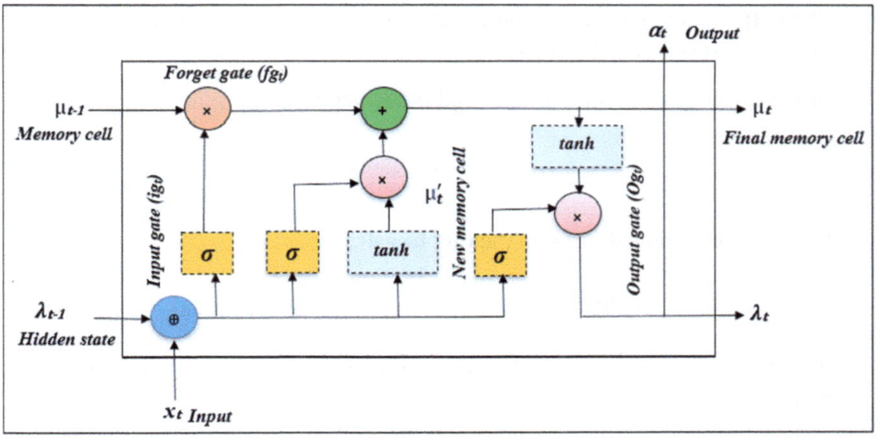

Fig. 4.4 Architecture of LSTM (modified and adapted from Vogeti et al., 2024 under CC BY-NC-ND 4.0 License)

ii. The information that passes through the input gate establishes the new information added further to the current memory state. At each time step, the algorithm combines the product of α_{ig} and λ_{t-1}, product of ω_{ig} and x_t and b_{ig}. Further, the activation function (σ) is applied to this combination (Eq. 4.10):

$$\text{Input gate}(ig_t) = \sigma(\omega_{ig}x_t + \alpha_{ig}\lambda_{t-1} + b_{ig}) \tag{4.10}$$

where ω_{ig}, α_{ig}, and b_{ig} are weight vectors at the current step, previous step, and bias vector, respectively, at the input gate.

iii. Further, the updation of memory cell states determines the information to be added to the memory state of an LSTM. This process involves the calculation of the new memory cell, μ_t', which is similar to the forget and input gates having weights (ω_μ, α_μ) and bias (b_μ) (Eqs. 4.10–4.11). Function tanh is employed for storing the information in the new memory cell μ_t' at t to facilitate a quicker convergence rate (Eq. 4.11):

$$\text{New memory cell}\ (\mu_t') = \tanh(\omega_\mu x_t + \alpha_\mu \lambda_{t-1} + b_\mu) \tag{4.11}$$

where ω_μ, α_μ, b_μ are weight vectors at the current step, previous step, and bias vector, respectively, at the cell.

Information passing through the final memory cell is (Eq. 4.12)

$$\mu_t = fg_t\mu_{t-1} + ig_t\mu_t' \tag{4.12}$$

Here, μ_{t-1} describes new memory cells at time $t - 1$

iv. The output gate manages the propagation of information from the recent cell to the final state (Eqs. 4.13–4.14):

Information passing through the final hidden cell:

$$\lambda_t = og_t \tanh(\mu_t) \tag{4.13}$$

Information obtained through the output gate:

$$\text{Output gate } (og_t) = \sigma(\omega_{og} x_t + \alpha_{og} \lambda_{t-1} + b_{og}) \tag{4.14}$$

where ω_{og}, α_{og}, and b_{og} are weight vectors at the current step, previous step, and bias vectors at the output gate.

Figure 4.5 shows the workflow of LSTM. Batch size, layer node, number of nodes, epochs, learning rate, and dropout exhibit their influence on the weights updated in the gating units. An increase in the values of the LSTM layer node, learning rate, and epochs positively affects the algorithm performance.

Numerical Problem 4.3. Rainfall, $x_t = (4,5,6)$ yields a runoff, $y_t = (1,1.5,2)$. Establish a relationship using LSTM with three hidden units $\lambda_{t-1} = (1,2,3)$. Assume the related weights appropriately. Refer to Fig. 4.5 for understanding the working steps of the problem.

Solution:

Given, $x_t = (4,5,6)$; $\lambda_{t-1} = (1,2,3)$

$$(\lambda_{t-1}, x_t) = (1, 2, 3, 4, 5, 6)$$

Assume the weights μ_{t-1} as (5,5,5).

Step 1 Random initialization of weights and biases for the gating units:

Input weight vector in forget gate $\omega_{fg} = \begin{bmatrix} 0 & 0 & 0 & 0 & 0 & -1 \\ 5 & 6 & 7 & 8 & 9 & 10 \\ 3 & 4 & 5 & 6 & 7 & 8 \end{bmatrix}$. Bias vector in forget gate

$b_{fg} = \begin{bmatrix} 1 \\ 2 \\ 3 \end{bmatrix}$.

Input weight vector in input gate $\omega_{ig} = \begin{bmatrix} 1 & 1 & 1 & 1 & 1 & 1 \\ 2 & 2 & 2 & 2 & 2 & 2 \\ 3 & 3 & 3 & 3 & 3 & 3 \end{bmatrix}$.

Bias vector in input gate $b_{ig} = \begin{bmatrix} 1 \\ 1 \\ 1 \end{bmatrix}$.

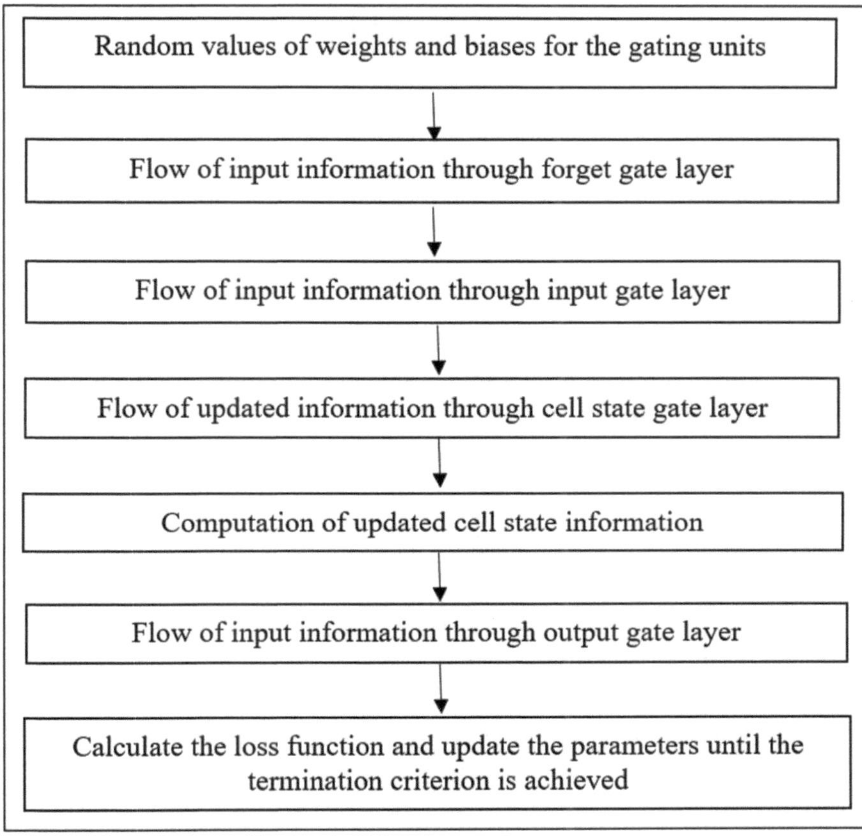

Fig. 4.5 Workflow of LSTM

Input weight vector in cell state $\omega_{\mu} = \begin{bmatrix} 1 & 1 & 1 & 1 & 1 & 1 \\ 2 & 2 & 2 & 2 & 2 & 2 \\ -3 & -3 & -3 & -3 & -3 & -3 \end{bmatrix}$

Bias vector in cell state $b_{\mu} = \begin{bmatrix} 1 \\ 1 \\ 1 \end{bmatrix}$

Input weight vector in output gate $\omega_{og} = \begin{bmatrix} 0.25 & 0.25 & 0.25 & 0.25 & 0.25 & 0.25 \\ 0.10 & 0.10 & 0.10 & 0.10 & 0.10 & 0.10 \\ 0.50 & 0.50 & 0.50 & 0.50 & 0.50 & 0.50 \end{bmatrix}$

Bias vector in output gate $b_{og} = \begin{bmatrix} 0.75 \\ 0.90 \\ 0.50 \end{bmatrix}$

$$(\lambda_{t-1}, x_t) = \begin{bmatrix} 1 \\ 2 \\ 3 \\ 4 \\ 5 \\ 6 \end{bmatrix}$$

Step 2 Flow of input information through forget gate layer $\mathrm{fg_t} = \sigma'\ (\omega_{\mathrm{fg}}.(\lambda_{t-1}, x_t) + b_{\mathrm{fg}})$:

$$\omega_{fg}(\lambda_{t-1}, x_t) = \begin{bmatrix} 0\,0\,0\,0\,0 & -1 \\ 5\,6\,7\,8\,9 & 10 \\ 3\,4\,5\,6\,7 & 8 \end{bmatrix} \cdot \begin{bmatrix} 1 \\ 2 \\ 3 \\ 4 \\ 5 \\ 6 \end{bmatrix}$$

$$= \begin{bmatrix} (0 \times 1) + (0 \times 2) + (0 \times 3) + (0 \times 4) + (0 \times 5) + (-1 \times 6) \\ (5 \times 1) + (6 \times 2) + (7 \times 3) + (8 \times 4) + (9 \times 5) + (10 \times 6) \\ (3 \times 1) + (4 \times 2) + (5 \times 3) + (6 \times 4) + (7 \times 5) + (8 \times 6) \end{bmatrix}$$

$$= \begin{bmatrix} 0+0+0+0+0-6 \\ 5+12+21+32+45+60 \\ 3+8+15+24+35+48 \end{bmatrix} = \begin{bmatrix} -6 \\ 175 \\ 133 \end{bmatrix}$$

$$\omega_{\mathrm{fg}}\big((\lambda_{t-1}, x_t) + b_{\mathrm{fg}}\big) = \begin{bmatrix} -6 \\ 175 \\ 133 \end{bmatrix} + \begin{bmatrix} 1 \\ 2 \\ 3 \end{bmatrix} = \begin{bmatrix} -5 \\ 177 \\ 136 \end{bmatrix}$$

$\mathrm{fg_t} = \sigma'(\omega_{\mathrm{fg}}.(\lambda_{t-1}, x_t) + b_{\mathrm{fg}})$ where, $\sigma'(x) = \frac{1}{1+e^{-x}}$

Flow of input information through the forget gate layer $\mathrm{fg_t} = \sigma'\left(\begin{bmatrix} -5 \\ 177 \\ 136 \end{bmatrix}\right)$

$$= \begin{bmatrix} \frac{1}{1+e^{-(-5)}} \\ \frac{1}{1+e^{-177}} \\ \frac{1}{1+e^{-136}} \end{bmatrix}$$

$$= \begin{bmatrix} 0.0069 \\ 1 \\ 1 \end{bmatrix} \quad \text{(i)}$$

Step 3 Flow of input information through the input gate layer $ig_t = \sigma'(\omega_{ig}.(\lambda_{t-1}, x_t) + b_{ig})$:

$$\omega_{ig}(\lambda_{t-1}, x_t) = \begin{bmatrix} 1\,1\,1\,1\,1\,1 \\ 2\,2\,2\,2\,2\,2 \\ 3\,3\,3\,3\,3\,3 \end{bmatrix} \cdot \begin{bmatrix} 1 \\ 2 \\ 3 \\ 4 \\ 5 \\ 6 \end{bmatrix}$$

$$= \begin{bmatrix} (1\times 1) + (1\times 2) + (1\times 3) + (1\times 4) + (1\times 5) + (1\times 6) \\ (2\times 1) + (2\times 2) + (2\times 3) + (2\times 4) + (2\times 5) + (2\times 6) \\ (3\times 1) + (3\times 2) + (3\times 3) + (3\times 4) + (3\times 5) + (3\times 6) \end{bmatrix}$$

$$= \begin{bmatrix} 21 \\ 42 \\ 63 \end{bmatrix}$$

$$\omega_{ig}\big((\lambda_{t-1}, x_t) + b_{ig}\big) = \begin{bmatrix} 21 \\ 42 \\ 63 \end{bmatrix} + \begin{bmatrix} 1 \\ 1 \\ 1 \end{bmatrix} = \begin{bmatrix} 22 \\ 43 \\ 64 \end{bmatrix}$$

$$ig_t = \sigma'\big(\omega_{ig}(\lambda_{t-1}, x_t) + b_{ig}\big) = \sigma'\left(\begin{bmatrix} 22 \\ 43 \\ 64 \end{bmatrix}\right)$$

$$= \begin{bmatrix} \frac{1}{1+e^{-22}} \\ \frac{1}{1+e^{-43}} \\ \frac{1}{1+e^{-64}} \end{bmatrix}$$

$$= \begin{bmatrix} 0.999 \\ 1 \\ 1 \end{bmatrix} \qquad \text{(ii)}$$

Step 4 Flow of updated information through cell state gate layer $\widetilde{\mu_t} = \tanh(\omega_{\mu}.(\lambda_{t-1}, x_t) + b_{\mu})$:

$$\omega_{\mu}(\lambda_{t-1}, x_t) = \begin{bmatrix} 1 & 1 & 1 & 1 & 1 & 1 \\ 2 & 2 & 2 & 2 & 2 & 2 \\ -3 & -3 & -3 & -3 & -3 & -3 \end{bmatrix} \cdot \begin{bmatrix} 1 \\ 2 \\ 3 \\ 4 \\ 5 \\ 6 \end{bmatrix}$$

$$= \begin{bmatrix} (1 \times 1) + (1 \times 2) + (1 \times 3) + (1 \times 4) + (1 \times 5) + (1 \times 6) \\ (2 \times 1) + (2 \times 2) + (2 \times 3) + (2 \times 4) + (2 \times 5) + (2 \times 6) \\ (-3 \times 1) + (-3 \times 2) + (-3 \times 3) + (-3 \times 4) + (-3 \times 5) + (-3 \times 6) \end{bmatrix}$$

$$= \begin{bmatrix} 21 \\ 42 \\ -63 \end{bmatrix}$$

$$\omega_\mu\left((\lambda_{t-1}, x_t) + b_\mu\right) = \begin{bmatrix} 21 \\ 42 \\ -63 \end{bmatrix} + \begin{bmatrix} 1 \\ 1 \\ 1 \end{bmatrix} = \begin{bmatrix} 22 \\ 43 \\ -62 \end{bmatrix}$$

$\widetilde{\mu_t} = \tanh(\omega_\mu.(\lambda_{t-1}, x_t) + b_\mu)$, where $\tanh(x) = \frac{e^x - e^{-x}}{e^x + e^{-x}}$

$$\widetilde{\mu_t} = \tanh\left(\begin{bmatrix} 22 \\ 43 \\ -62 \end{bmatrix}\right) = \begin{bmatrix} \frac{e^{22}-e^{-22}}{e^{22}+e^{-22}} \\ \frac{e^{43}-e^{-43}}{e^{43}+e^{-43}} \\ \frac{e^{-62}-e^{-(-62)}}{e^{-62}+e^{-(-62)}} \end{bmatrix} = \begin{bmatrix} 1 \\ 1 \\ -1 \end{bmatrix} \quad \text{(iii)}$$

Step 5 Computation of updated cell state information $\mu_t = fg_t\mu_{t-1} + ig_t\widetilde{\mu_t}$

Substitute Eqs. (i), (ii), and (iii) for computing the final updated cell state information:

$$\mu_t = fg_t\mu_{t-1} + ig_t\widetilde{\mu_t} = \begin{bmatrix} 0.0069 \\ 1 \\ 1 \end{bmatrix} \cdot \begin{bmatrix} 5 \\ 5 \\ 5 \end{bmatrix} + \begin{bmatrix} 0.999 \\ 1 \\ 1 \end{bmatrix} \cdot \begin{bmatrix} 1 \\ 1 \\ -1 \end{bmatrix}$$

$$= \begin{bmatrix} 1.0325 \\ 6 \\ 4 \end{bmatrix} \quad \text{(iv)}$$

Step 6 Flow of input information through the output gate layer $Og_t = \sigma'(\omega_{og}.(\lambda_{t-1}, x_t) + b_{og})$:

$$\omega_{og}(\lambda_{t-1}, x_t) = \begin{bmatrix} 0.25 & 0.25 & 0.25 & 0.25 & 0.25 & 0.25 \\ 0.10 & 0.10 & 0.10 & 0.10 & 0.10 & 0.10 \\ 0.50 & 0.50 & 0.50 & 0.50 & 0.50 & 0.50 \end{bmatrix} \cdot \begin{bmatrix} 1 \\ 2 \\ 3 \\ 4 \\ 5 \\ 6 \end{bmatrix}$$

$$= \begin{bmatrix} (0.25 \times 1) + (0.25 \times 2) + (0.25 \times 3) + (0.25 \times 4) + (0.25 \times 5) + (0.25 \times 6) \\ (0.10 \times 1) + (0.10 \times 2) + (0.10 \times 3) + (0.10 \times 4) + (0.10 \times 5) + (0.10 \times 6) \\ (0.50 \times 1) + (0.50 \times 2) + (0.50 \times 3) + (0.50 \times 4) + (0.50 \times 5) + (0.50 \times 6) \end{bmatrix}$$

$$= \begin{bmatrix} 5.25 \\ 2.10 \\ 10.50 \end{bmatrix}$$

$$\omega_{og}(\lambda_{t-1}, x_t) + b_{og} = \begin{bmatrix} 5.25 \\ 2.10 \\ 10.50 \end{bmatrix} + \begin{bmatrix} 0.75 \\ 0.90 \\ 0.50 \end{bmatrix} = \begin{bmatrix} 6 \\ 3 \\ 11 \end{bmatrix}$$

$$Og_t = \sigma'\left(\omega_{og}(\lambda_{t-1}, x_t) + b_{og}\right) = \sigma'\left(\begin{bmatrix} 6 \\ 3 \\ 11 \end{bmatrix}\right)$$

$$= \begin{bmatrix} \frac{1}{1+e^{-6}} \\ \frac{1}{1+e^{-3}} \\ \frac{1}{1+e^{-11}} \end{bmatrix}$$

$$= \begin{bmatrix} 0.9975 \\ 0.9526 \\ 0.9999 \end{bmatrix} \quad \text{(v)}$$

Step 7 Computation of hidden state information $\lambda_t = Og_t \tanh(\mu_t)$

Substitute Eqs. (iv) and (v) to calculate the hidden state information:

$$\lambda_t = \begin{bmatrix} 0.9975 \\ 0.9526 \\ 0.9999 \end{bmatrix} \tanh\left(\begin{bmatrix} 1.0325 \\ 6 \\ 4 \end{bmatrix}\right) = \begin{bmatrix} 0.9975 \\ 0.9526 \\ 0.9999 \end{bmatrix} \cdot \begin{bmatrix} \frac{e^{1.0325}-e^{-1.0325}}{e^{1.0325}+e^{-1.0325}} \\ \frac{e^{6}-e^{-6}}{e^{6}+e^{-6}} \\ \frac{e^{4}-e^{-4}}{e^{4}+e^{-4}} \end{bmatrix}$$

$$= \begin{bmatrix} 0.9975 \\ 0.9526 \\ 0.9999 \end{bmatrix} \cdot \begin{bmatrix} 0.775 \\ 0.9999 \\ 0.9993 \end{bmatrix} = \begin{bmatrix} 0.773 \\ 0.9525 \\ 0.9992 \end{bmatrix}$$

Step 8 Calculate the loss function:

The predicted outputs are compared with the observed, and the loss function is computed (Table 4.1).

Table 4.1 Loss function

Observed (y_t)	Predicted (h_t)	Loss function
1	0.773	0.227
1.5	0.9525	0.5475
2	0.9992	1.0008

Numerical Problem 4.4. Relate building strength, $x_t = (0.5,0.75,1)$ and vulnerability to earthquakes, $y_t = (0.1,0.25,0.4)$ using LSTM. Three hidden units $\lambda_{t-1} = (0.01,0.04,0.09)$ are suggested. Assume the related weights appropriately. Refer to Fig. 4.5 for understanding the working steps of the problem.

Solution:

Given, $x_t = (0.5,0.75,1)$; $\lambda_{t-1} = (0.01, 0.04, 0.09)$

$(\lambda_{t-1}, x_t) = (0.01, 0.04, 0.09, 0.5, 0.75, 1)$

Assume the weights μ_{t-1} as (3,3,3)

Step 1 Random initialization of weights and biases for the gating units:

$$\text{Input weight vector in forget gate } \omega_{fg} = \begin{bmatrix} 0 & 0 & 0 & 0 & 0 & -5 \\ 3 & 2 & 1 & 0 & 4 & 10 \\ 4 & 7 & 5 & 1 & 7 & 15 \end{bmatrix}.$$

$$\text{Bias vector in forget gate } b_{fg} = \begin{bmatrix} 1 \\ 1 \\ 1 \end{bmatrix}.$$

$$\text{Input weight vector in input gate } \omega_{ig} = \begin{bmatrix} 1 & 1 & 1 & 1 & 1 & 1 \\ 1 & 1 & 1 & 1 & 1 & 1 \\ 1 & 1 & 1 & 1 & 1 & 1 \end{bmatrix}.$$

$$\text{Bias vector in input gate } b_{ig} = \begin{bmatrix} 1 \\ 0.5 \\ 1 \end{bmatrix}.$$

$$\text{Input weight vector in cell state } \omega_{\mu} = \begin{bmatrix} 1 & 1 & 1 & 1 & 1 & 1 \\ 0.5 & 0.5 & 0.5 & 0.5 & 0.5 & 0.5 \\ -1 & -1 & -1 & -1 & -1 & -1 \end{bmatrix}$$

$$\text{Bias vector in cell state } b_{\mu} = \begin{bmatrix} 0.5 \\ 0.5 \\ 0.5 \end{bmatrix}$$

$$\text{Input weight vector in output gate } \omega_{og} = \begin{bmatrix} 0.75 & 0.75 & 0.75 & 0.75 & 0.75 & 0.75 \\ 0.20 & 0.20 & 0.20 & 0.20 & 0.20 & 0.20 \\ 1 & 1 & 1 & 1 & 1 & 1 \end{bmatrix}$$

$$\text{Bias vector in output gate } b_{og} = \begin{bmatrix} 0.65 \\ 0.70 \\ 0.40 \end{bmatrix}$$

$$(\lambda_{t-1}, x_t) = \begin{bmatrix} 0.01 \\ 0.04 \\ 0.09 \\ 0.5 \\ 0.75 \\ 1 \end{bmatrix}$$

Step 2 Flow of input information through forget gate layer $fg_t = \sigma'(\omega_{fg}.(\lambda_{t-1}, x_t) + b_{fg})$:

$$\omega_{fg}(\lambda_{t-1}, x_t) = \begin{bmatrix} 0 & 0 & 0 & 0 & 0 & -5 \\ 3 & 2 & 1 & 0 & 4 & 10 \\ 4 & 7 & 5 & 1 & 7 & 15 \end{bmatrix} \cdot \begin{bmatrix} 0.01 \\ 0.04 \\ 0.09 \\ 0.5 \\ 0.75 \\ 1 \end{bmatrix}$$

$$= \begin{bmatrix} (0 \times 0.01) + (0 \times 0.04) + (0 \times 0.09) + (0 \times 0.5) + (0 \times 0.75) + (-5 \times 1) \\ (3 \times 0.01) + (2 \times 0.04) + (1 \times 0.09) + (0 \times 0.5) + (4 \times 0.75) + (10 \times 1) \\ (4 \times 0.01) + (7 \times 0.04) + (5 \times 0.09) + (1 \times 0.5) + (7 \times 0.75) + (15 \times 1) \end{bmatrix}$$

$$= \begin{bmatrix} -5 \\ 13.2 \\ 21.52 \end{bmatrix}$$

$$\left(\omega_{fg}(\lambda_{t-1}, x_t) + b_{fg}\right) = \begin{bmatrix} -5 \\ 13.2 \\ 21.52 \end{bmatrix} + \begin{bmatrix} 1 \\ 1 \\ 1 \end{bmatrix} = \begin{bmatrix} -4 \\ 14.2 \\ 22.52 \end{bmatrix}$$

$$fg_t = \sigma'\left(\omega_{fg}(\lambda_{t-1}, x_t) + b_{fg}\right) = \sigma'\left(\begin{bmatrix} -4 \\ 14.2 \\ 22.52 \end{bmatrix}\right)$$

$$= \begin{bmatrix} \frac{1}{1+e^{-(-4)}} \\ \frac{1}{1+e^{-14.2}} \\ \frac{1}{1+e^{-22.52}} \end{bmatrix}$$

$$= \begin{bmatrix} 0.01799 \\ 1 \\ 1 \end{bmatrix} \qquad \text{(i)}$$

Step 3 Flow of input information through the input gate layer $ig_t = \sigma'(\omega_{ig}.(\lambda_{t-1}, x_t) + b_{ig})$:

$$\omega_{ig}(\lambda_{t-1}, x_t) = \begin{bmatrix} 1 & 1 & 1 & 1 & 1 & 1 \\ 1 & 1 & 1 & 1 & 1 & 1 \\ 1 & 1 & 1 & 1 & 1 & 1 \end{bmatrix} \cdot \begin{bmatrix} 0.01 \\ 0.04 \\ 0.09 \\ 0.5 \\ 0.75 \\ 1 \end{bmatrix}$$

$$= \begin{bmatrix} (1 \times 0.01) + (1 \times 0.04) + (1 \times 0.09) + (1 \times 0.5) + (1 \times 0.75) + (1 \times 1) \\ (1 \times 0.01) + (1 \times 0.04) + (1 \times 0.09) + (1 \times 0.5) + (1 \times 0.75) + (1 \times 1) \\ (1 \times 0.01) + (1 \times 0.04) + (1 \times 0.09) + (1 \times 0.5) + (1 \times 0.75) + (1 \times 1) \end{bmatrix}$$

$$= \begin{bmatrix} 2.39 \\ 2.39 \\ 2.39 \end{bmatrix}$$

$$\left(\omega_{ig}(\lambda_{t-1}, x_t) + b_{ig}\right) = \begin{bmatrix} 2.39 \\ 2.39 \\ 2.39 \end{bmatrix} + \begin{bmatrix} 1 \\ 0.5 \\ 1 \end{bmatrix} = \begin{bmatrix} 3.39 \\ 2.89 \\ 3.39 \end{bmatrix}$$

$$ig_t = \sigma'\left(\omega_{ig}(\lambda_{t-1}, x_t) + b_{ig}\right) = \sigma'\left(\begin{bmatrix} 3.39 \\ 2.89 \\ 3.39 \end{bmatrix}\right)$$

$$= \begin{bmatrix} \frac{1}{1+e^{-3.39}} \\ \frac{1}{1+e^{-2.89}} \\ \frac{1}{1+e^{-3.39}} \end{bmatrix}$$

$$= \begin{bmatrix} 0.9674 \\ 0.9473 \\ 0.9674 \end{bmatrix} \quad \text{(ii)}$$

Step 4 Flow of updated information through cell state gate layer $\widetilde{\mu_t} = \tanh(\omega_{\mu}.(\lambda_{t-1}, x_t) + b_{\mu})$:

$$\omega_{\mu}(\lambda_{t-1}, x_t) = \begin{bmatrix} 1 & 1 & 1 & 1 & 1 & 1 \\ 0.5 & 0.5 & 0.5 & 0.5 & 0.5 & 0.5 \\ -1 & -1 & -1 & -1 & -1 & -1 \end{bmatrix} \cdot \begin{bmatrix} 0.01 \\ 0.04 \\ 0.09 \\ 0.5 \\ 0.75 \\ 1 \end{bmatrix}$$

$$= \begin{bmatrix} (1 \times 0.01) + (1 \times 0.04) + (1 \times 0.09) + (1 \times 0.5) + (1 \times 0.75) + (1 \times 1) \\ (0.5 \times 0.01) + (0.5 \times 0.04) + (0.5 \times 0.09) + (0.5 \times 0.5) + (0.5 \times 0.75) + (0.5 \times 1) \\ (-1 \times 0.01) + (-1 \times 0.04) + (-1 \times 0.09) + (-1 \times 0.5) + (-1 \times 0.75) + (-1 \times 1) \end{bmatrix}$$

$$= \begin{bmatrix} 2.39 \\ 1.195 \\ -2.39 \end{bmatrix}$$

$$\left(\omega_{\mu}(\lambda_{t-1}, x_t) + b_{\mu}\right) = \begin{bmatrix} 2.39 \\ 1.195 \\ -2.39 \end{bmatrix} + \begin{bmatrix} 0.5 \\ 0.5 \\ 0.5 \end{bmatrix} = \begin{bmatrix} 2.89 \\ 1.695 \\ -1.89 \end{bmatrix}$$

$$\widetilde{\mu_t} = \tanh\left(\omega_{\mu}(\lambda_{t-1}, x_t) = b_{\mu}\right)$$

$$\widetilde{\mu_t} = \tanh\left(\begin{bmatrix} 2.89 \\ 1.695 \\ -1.89 \end{bmatrix}\right) = \begin{bmatrix} \frac{e^{2.89} - e^{-2.89}}{e^{2.89} + e^{-2.89}} \\ \frac{e^{1.695} - e^{-1.695}}{e^{1.695} + e^{-1.695}} \\ \frac{e^{-1.89} - e^{-(-1.89)}}{e^{-1.89} + e^{-(-1.89)}} \end{bmatrix}$$

$$= \begin{bmatrix} 0.943 \\ 0.805 \\ 2.97722 \times 10^{-5} \end{bmatrix} \qquad \text{(iii)}$$

Step 5 Computation of updated cell state information $\mu_t = fg_t\mu_{t-1} + ig_t\widetilde{\mu_t}$.

Substitute Eqs. (i), (ii), and (iii) for computing the final updated cell state information:

$$\mu_t = fg_t\mu_{t-1} + ig_t\widetilde{\mu_t} = \begin{bmatrix} 0.01799 \\ 1 \\ 1 \end{bmatrix} \cdot \begin{bmatrix} 3 \\ 3 \\ 3 \end{bmatrix} + \begin{bmatrix} 0.9674 \\ 0.9473 \\ 0.9674 \end{bmatrix} \cdot \begin{bmatrix} 0.943 \\ 0.805 \\ 2.97722 \times 10^{-5} \end{bmatrix}$$

$$\mu_t = \begin{bmatrix} 0.1005 \\ 3 \\ 3 \end{bmatrix} + \begin{bmatrix} 0.9123 \\ 0.7626 \\ 2.8802 \times 10^{-5} \end{bmatrix} = \begin{bmatrix} 1.0128 \\ 3.7626 \\ 3 \end{bmatrix} \qquad \text{(iv)}$$

Step 6 Flow of input information through the output gate layer $Og_t = \sigma'(\omega_{og}.(\lambda_{t-1}, x_t) + b_{og})$:

$$\omega_{og}(\lambda_{t-1}, x_t) = \begin{bmatrix} 0.75 & 0.75 & 0.75 & 0.75 & 0.75 & 0.75 \\ 0.20 & 0.20 & 0.20 & 0.20 & 0.20 & 0.20 \\ 1 & 1 & 1 & 1 & 1 & 1 \end{bmatrix} \cdot \begin{bmatrix} 0.01 \\ 0.04 \\ 0.09 \\ 0.5 \\ 0.75 \\ 1 \end{bmatrix}$$

$$= \begin{bmatrix} (0.75 \times 0.01) + (0.75 \times 0.04) + (0.75 \times 0.09) + (0.75 \times 0.5) + (0.75 \times 0.75) + (0.75 \times 1) \\ (0.20 \times 0.01) + (0.20 \times 0.04) + (0.20 \times 0.09) + (0.20 \times 0.5) + (0.20 \times 0.75) + (0.20 \times 1) \\ (1 \times 0.01) + (1 \times 0.04) + (1 \times 0.09) + (1 \times 0.5) + (1 \times 0.75) + (1 \times 1) \end{bmatrix}$$

$$= \begin{bmatrix} 1.7925 \\ 0.478 \\ 2.39 \end{bmatrix}$$

$$\omega_{og}(\lambda_{t-1}, x_t) + b_{og} = \begin{bmatrix} 1.7925 \\ 0.478 \\ 2.39 \end{bmatrix} + \begin{bmatrix} 0.65 \\ 0.70 \\ 0.40 \end{bmatrix} = \begin{bmatrix} 2.4425 \\ 1.178 \\ 2.79 \end{bmatrix}$$

$$\begin{aligned} Og_t = \sigma'\left(\omega_{og}(\lambda_{t-1}, x_t) + b_{og}\right) &= \sigma'\left(\begin{bmatrix} 2.4425 \\ 1.178 \\ 2.79 \end{bmatrix}\right) \\ &= \begin{bmatrix} \frac{1}{1+e^{-2.4425}} \\ \frac{1}{1+e^{-1.178}} \\ \frac{1}{1+e^{-2.79}} \end{bmatrix} \\ &= \begin{bmatrix} 0.92 \\ 0.7646 \\ 0.942 \end{bmatrix} \end{aligned} \tag{v}$$

Step 7 Computation of hidden state information $\lambda_t = Og_t \tanh(\mu_t)$:

Substitute Eqs. (iv) and (v) to calculate the hidden state information:

$$\begin{aligned} \lambda_t &= \begin{bmatrix} 0.92 \\ 0.7646 \\ 0.942 \end{bmatrix} \tanh\left(\begin{bmatrix} 1.0128 \\ 3.7626 \\ 3 \end{bmatrix}\right) = \begin{bmatrix} 0.92 \\ 0.7646 \\ 0.942 \end{bmatrix} \cdot \begin{bmatrix} \frac{e^{1.0128}-e^{-1.0128}}{e^{1.0128}+e^{-1.0128}} \\ \frac{e^{3.7626}-e^{-3.7626}}{e^{3.7626}+e^{-3.7626}} \\ \frac{e^{3}-e^{-3}}{e^{3}+e^{-3}} \end{bmatrix} \\ &= \begin{bmatrix} 0.92 \\ 0.7646 \\ 0.942 \end{bmatrix} \cdot \begin{bmatrix} 0.767 \\ 0.9989 \\ 0.995 \end{bmatrix} = \begin{bmatrix} 0.7056 \\ 0.7638 \\ 0.9373 \end{bmatrix} \end{aligned}$$

Step 8 Calculate the loss function:

The predicted outputs are compared with the observed, and the loss function is computed (Table 4.2).

Table 4.2 Loss function

Observed (y_t)	Predicted (h_t)	Loss function
0.1	0.7056	0.6056
0.25	0.7638	0.5138
0.4	0.9373	0.5373

4.5 Bi-Directional-LSTM

Bi-LSTM (Fig. 4.6) is an improvised variation of LSTM competent in capturing the past and future of a time series (Roy et al., 2022). It includes two LSTMs to accomplish the forward and backward computations of the hidden vectors, $\overrightarrow{\lambda}_t$ and $\overleftarrow{\lambda}_t$, respectively (Eqs. 4.15 and 4.16).

$$\overrightarrow{\lambda}_t = f\left(\omega_1 x_t + \omega_2 \overrightarrow{\lambda}_{t-1}\right) \tag{4.15}$$

$$\overleftarrow{\lambda}_t = f\left(\omega_3 x_t + \omega_5 \overleftarrow{\lambda}_{t-1}\right) \tag{4.16}$$

The average outputs from both LSTMs are the basis for the forecast, og_t (Eq. 4.17):

$$og_t = f\left(\omega_4 \overrightarrow{\lambda}_t + \omega_6 \overleftarrow{\lambda}_t + \text{Bias}\right) \tag{4.17}$$

where ω_1 is the weight for input to the forward layer; ω_3 is the weight for input to the backward layer; ω_2, ω_5 denote the weights for hidden-to-hidden layers; ω_4 is the

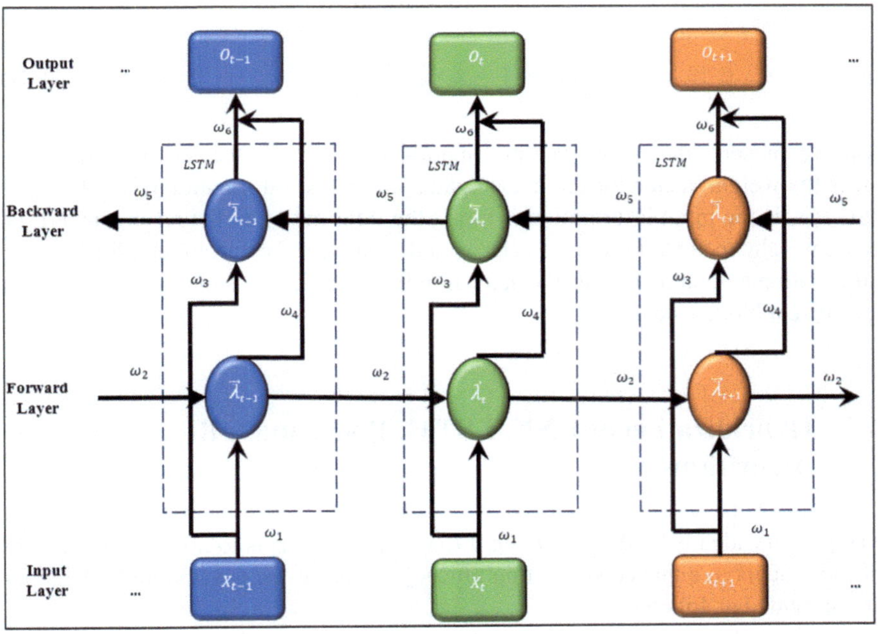

Fig. 4.6 Architecture of Bi-LSTM (adapted from Deb et al., 2024 under CC BY 4.0 License)

weight for forwarding to the output layer; and ω_6 is the weight for backward to the output layer (Deb et al., 2024).

4.6 Gated Recurrent Unit

GRU (refer to Fig. 4.7) has a relatively simplified architecture compared to LSTM, resulting in fewer parameters and operations and making them faster to train (Shen et al., 2018). It is based on updating and reset gates. The update gate helps decide the amount of information that needs to be retrieved from the past information. On the contrary, the reset gate establishes the basis for the amount of information that needs to be forgotten. The helpful information is finally stored with the help of current and final hidden states. The mathematical expressions for GRU are presented in Eqs. (4.18–4.21):

$$z_t = \sigma(\omega_z \cdot (\lambda_{t-1}, x_t)) \tag{4.18}$$

$$r_t = \sigma(\omega_r \cdot (\lambda_{t-1}, x_t)) \tag{4.19}$$

$$\lambda_t = \tanh(\omega.(r_t\lambda_{t-1}, x_t)) \tag{4.20}$$

$$\widetilde{\lambda_t} = (1 - z_t)\lambda_{t-1} + (z_t\lambda_t) \tag{4.21}$$

where z_t presents the update gate information at time t; σ is the activation function; ω_z is the weight vector for the update gate; λ_{t-1} is the information of hidden state at t–1; x_t is the input information at t; r_t is the information of the reset gate at t; ω_r is the weight vector for the reset gate; λ_t and $\widetilde{\lambda_t}$ present the current and final hidden information at t; tanh presents the activation function; and ω is the weight vector for the current hidden state.

4.7 Hybridization of CNN, LSTM, RNN, and GRU Algorithms

The primary idea behind hybridizing different algorithms is to utilize the strengths of individuals to enhance simulating efficacy. Some of the possible hybridizations are presented as follows:

The CNN-LSTM utilizes the strengths of CNN and LSTM to enhance simulating ability. CNN is efficient at seizing high-dimensional spatial features of data (LSTM does not have this capability) with the assistance of convolution filters. It is less

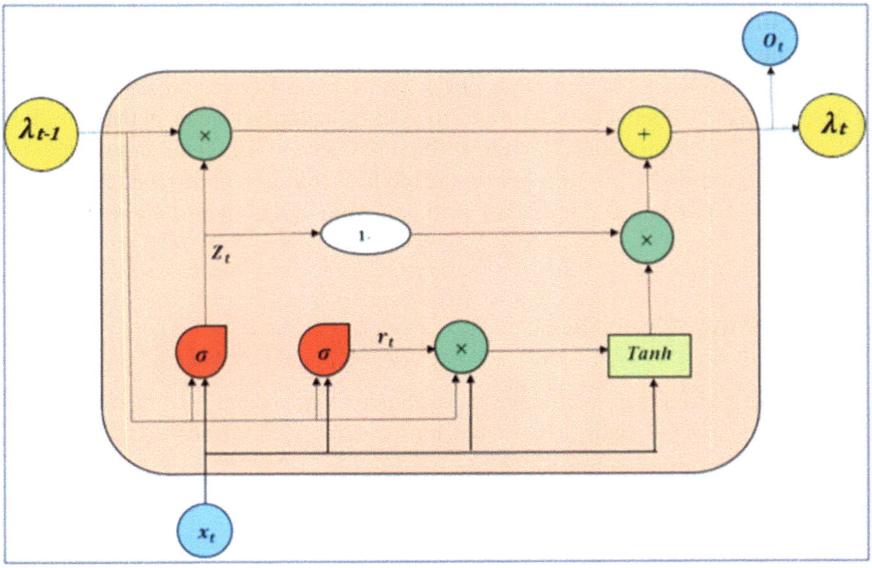

Fig. 4.7 Architecture of GRU

competent to establish long-term temporal dependencies (LSTM has this capability, which can be facilitated by memory and gating units). In this context, CNN-LSTM architecture can efficiently consider spatiotemporal details of data.

The GRU-RNN tackles the challenge of vanishing gradient found in RNNs. They can handle long-term interdependence in time series data by selectively updating and resetting information. GRU output is fed into the RNN algorithm as input. The output obtained from the RNN is the outcome of the hybrid algorithm.

The GRU-LSTM architecture combines GRU and LSTM units to capture and process temporal dependencies in sequential data. The architecture begins with an input layer that receives sequential data fed into the GRU layer. Further, the output of GRU is fed into the LSTM architecture, which further passes through three gates and generates output.

The RNN-LSTM algorithm is developed to handle mid- and long-term dependencies of sequential data efficiently. RNN output is fed into the LSTM algorithm as input. The output from LSTM can be considered as the outcome of the hybrid algorithm.

The CNN-GRU efficiently handles both spatial and temporal dependencies of sequential data. This architecture starts with a CNN layer to bring spatial features from the input data. Further, this output is fed as input to the GRU layer. The output from GRU can be considered as the outcome of the hybrid algorithm.

The CNN-GRU-LSTM architecture works similar to the mechanism of CNN-GRU. It only differs with an LSTM layer added to CNN-GRU. The output obtained from CNN-GRU is further fed into the LSTM layer, and output from this is considered the outcome of the hybrid algorithm.

The GRU-RNN-LSTM combines GRU with typical RNNs and LSTMs. GRU, RNN, and LSTM effectively learn short- to long-term dependencies in the data (Ren et al., 2022). The input data is first transmitted to GRU, which generates output and is then fed into the RNN. Furthermore, the RNN output is input to the LSTM. The output from LSTM can be considered as the outcome of the hybrid algorithm.

4.8 Boosting Algorithms

The principle behind these techniques is to construct an ensemble of Decision Trees (DTs) to decrease the error. Three algorithms, namely, AdaBoost, XGBoost, and CatBoost are described in this section.

4.8.1 *Adaptive Boosting*

AdaBoost selects features to improve algorithm prediction. It makes an ensemble out of weak learners to increase performance (Aldrees et al., 2022; Ding et al., 2022). The hardness of each training dataset is given as input such that newly constructed trees group the tougher ones. Stump is one component in DT, and it is comprised of one node and two leaves. Some of these stumps have a more significant weightage in predicting data. Every succeeding stump corrects the errors of the previous stump. Lastly, the prediction is (Eqs. 4.22–4.23)

$$H(x) = \text{sign}\left(\sum_{t=1}^{T} \alpha_t h_t(x))\right) \tag{4.22}$$

where

$$\alpha_t = \frac{1}{2}\text{Log}\left(\frac{1-\varepsilon_t}{\varepsilon_t}\right) \tag{4.23}$$

where $h_t(x)$ is the prediction by tth weak classifier, $\alpha_{t \text{ and}}$ ε_t are the weight and fractions of misclassifications by the tth classifier. Here, the Log represents the natural logarithm.

Each sample is given a weight, $D_t(i)$. These are assumed to be equal initially, whose sum is unity. After each iteration, this sample weight is updated based on which samples were incorrectly classified. These are given a higher weight, which dictates that they are more likely to appear in the next iteration of the bootstrapped dataset. The bootstrap method estimates quantities regarding populations by averaging estimates from smaller data samples. Specifically, the samples are created by selecting observations from a larger data sample and choosing them repeatedly. This way, a given observation can appear in a sample more than once. The feature that distinguishes the classes is selected as the first stump. Accordingly, the weightage of the stump is established on its accuracy. Classified sample weights are (Eq. 4.24)

$$D_{t+1}(i) = \frac{D_t(i)e^{-\alpha_t y_i h_t(x_i)}}{\sum_i D_t(i)e^{-\alpha_t y_i h_t(x_i)}} \tag{4.24}$$

Then, a newer dataset of identical size as the original is randomly chosen by repeating samples from the preceding dataset. Higher weightage samples will be selected more often. Then, another iteration is carried out on this bootstrapped dataset, and so on (Madhuri et al., 2021; Mishra et al., 2024).

Numerical Problem 4.5. Classify the datasets presented in Table 4.3 using AdaBoost. Distance to Nearest Stream (DNS) and Evapotranspiration (ET) are features considered. Consider Gini impurity as an attribute selection measure.

Solution:

An AdaBoost tree is created in much the same way a DT is made. The difference is that AdaBoost uses stumps instead of full-blown DTs with several leaves. Consider the stump as an example. After splitting the data based on a given value, how well the condition splits the different data classes is noticed: positive class (+1) and negative class (−1). Here (Fig. 4.8), DNS of 5.2 as a divider correctly classifies five out of six flooded points [*here, points and datasets are used interchangeably*] and four out

Table 4.3 Dataset for the numerical problem

Dataset	DNS	ET	Did flood occur?	Observed y_i	In terms of AdaBoost terminology
1	2.4	10.2	No	0	−1
2	12	5.4	Yes	1	1
3	4.5	16	No	0	−1
4	7.6	20.3	Yes	1	1
5	9.3	14.5	Yes	1	1
6	4.9	7.8	Yes	1	1
7	8.1	14.2	No	0	−1
8	4.3	4.5	Yes	1	1
9	3.2	12.4	No	0	−1
10	5.5	5.5	Yes	1	1
11	7.2	11.2	Yes	1	1
12	4.5	8.5	No	0	−1

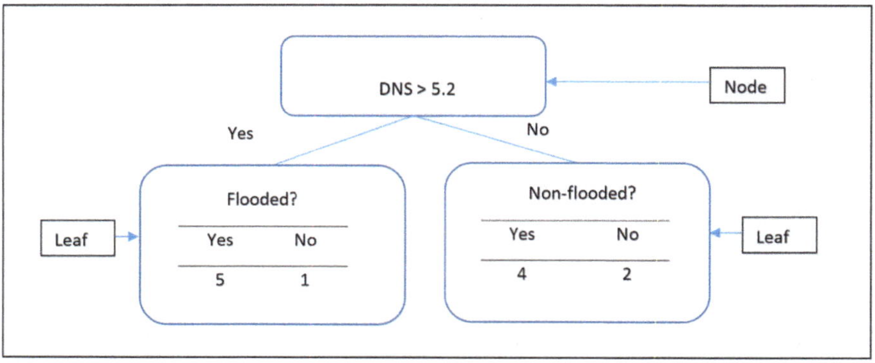

Fig. 4.8 Tree constructed using DNS at a value of 5.2

of six non-flooded points. In the left leaf, corresponding to Yes, it can be seen that five training examples are flooded, and one is not inundated despite being classified as flooded.

Similarly, on the right leaf, four out of six non-flooded points are correctly classified as Yes, and two are incorrectly classified as No. Many such dividers can be examined to pick the one that best splits the classes. A node refers to collecting all the datasets, and the leaves are the two groups of datasets formed based on a comparison.

Before the numerical problem begins, a recap of Gini Impurity is necessary. It is calculated to determine how well a specific node in a DT differentiates between the classes (flooded and non-flooded). For a given leaf, it is calculated as (Eq. 4.25)

$$\text{Gini impurity} = \left(1 - (\text{fraction of positive examples})^2 - (\text{fraction of negative examples})^2\right) \tag{4.25}$$

Suppose a leaf has eight training examples that belong to the positive class and five training examples that belong to the negative class. In this case, the Gini impurity index is calculated as

$$\text{Gini impurity} = \left(1 - \left(\frac{8}{5+8}\right)^2 - \left(\frac{5}{5+8}\right)^2\right) = 0.473$$

The weighted average of both leaves is taken to calculate the Gini impurity for a node and is performed to captivate the splitting efficiency of both leaves. If both leaves do an excellent job, their respective Gini impurities will be lower. Before constructing the first stump, each training example is assigned an equal weight, and the sum of the weights of all the training samples is 1 and can be placed next to the training data as a new column (Table 4.4).

The sample weight indicates how likely a given training example will be selected for the next tree that will be constructed; it does not impact the first tree built and

Table 4.4 Initial sample weights and classes of the entire dataset

Dataset	DNS	ET	Flooded (1)/non-flooded (−1) y_i	Sample weight
1	2.4	10.2	−1	$\frac{1}{12}$
2	12	5.4	1	$\frac{1}{12}$
3	4.5	16	−1	$\frac{1}{12}$
4	7.6	20.3	1	$\frac{1}{12}$
5	9.3	14.5	1	$\frac{1}{12}$
6	4.9	7.8	1	$\frac{1}{12}$
7	8.1	14.2	−1	$\frac{1}{12}$
8	4.3	4.5	1	$\frac{1}{12}$
9	3.2	12.4	−1	$\frac{1}{12}$
10	5.5	5.5	1	$\frac{1}{12}$
11	7.2	11.2	1	$\frac{1}{12}$
12	4.5	8.5	−1	$\frac{1}{12}$

serves as a starting point. To begin the construction of the first stump, find out which of the two features, DNS or ET, separates the training example better. To do this, evaluate the Gini impurity of both features. Before that, decide at which point each feature will give the highest possible Gini impurity. To make this possible, arrange the values for each feature in ascending order and consider splitting the data at the mean of each pair of consecutive training examples. The means of each pair of consecutive training examples are given in the midpoint column, and the datasets are split using each value present in this column (Table 4.5).

Table 4.5 DNS in ascending order

DNS	y_i	Midpoint
2.4	−1	
3.2	−1	2.8
4.3	1	3.75
4.5	−1	4.4
4.5	−1	4.5
4.9	1	4.7
5.5	1	5.2
7.2	1	6.35
7.6	1	7.4
8.1	−1	7.85
9.3	1	8.7
12	1	10.65

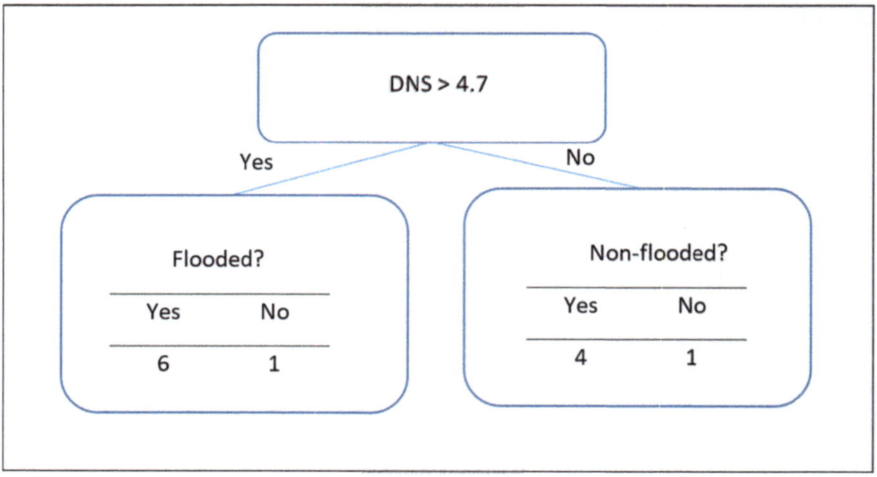

Fig. 4.9 Tree constructed using midpoint DNS at a value of 4.7

Take the example of the split made at midpoint 4.7 (refer to Fig. 4.9). There are five points with a DNS of less than 4.7 and seven points with a DNS greater than or equal to 4.7. The Gini impurity for each leaf can be computed.

$$\text{Gini impurity}_{\text{left}} = \left(1 - \left(\frac{6}{7}\right)^2 - \left(\frac{1}{7}\right)^2\right) = 0.245$$

$$\text{Gini impurity}_{\text{right}} = \left(1 - \left(\frac{1}{5}\right)^2 - \left(\frac{4}{5}\right)^2\right) = 0.320$$

The Gini impurity for the node is the weighted average of the Gini impurities of both leaves.

$$\text{Gini impurity}_{\text{DNS}=4.7} = \frac{0.245 \times 7 + 0.320 \times 5}{12} = 0.276$$

Similarly, the Gini impurities are calculated at each midpoint for DNS (Table 4.6).

Midpoints 4.5 and 4.7 achieve the same lowest Gini impurity of 0.276. So, the Gini impurity corresponding to DNS is 0.276. The process is repeated for ET (refer to Fig. 4.10, Table 4.7):

Adding the calculation for the last row in Table 4.7 for clarity:

The split is done at the ET value of 8.15 (vide Table 4.7).

$$\text{Gini impurity}_{\text{left}} = \left(1 - \left(\frac{1}{1}\right)^2\right) = 0$$

Table 4.6 DNS in ascending order—Gini impurity

DNS	y_i	Midpoint	Gini left	Gini right	Gini impurity
2.4	−1				
3.2	−1	2.8	0.463	0.000	0.424
4.3	1	3.75	0.420	0.000	0.350
4.5	−1	4.4	0.444	0.444	0.444
4.5	−1	4.5	0.245	0.320	**0.276**
4.9	1	4.7	0.245	0.320	**0.276**
5.5	1	5.2	0.278	0.444	0.361
7.2	1	6.35	0.320	0.490	0.419
7.6	1	7.4	0.375	0.500	0.458
8.1	−1	7.85	0.444	0.494	0.481
9.3	1	8.7	0.000	0.500	0.417
12	1	10.65	0.000	0.496	0.455

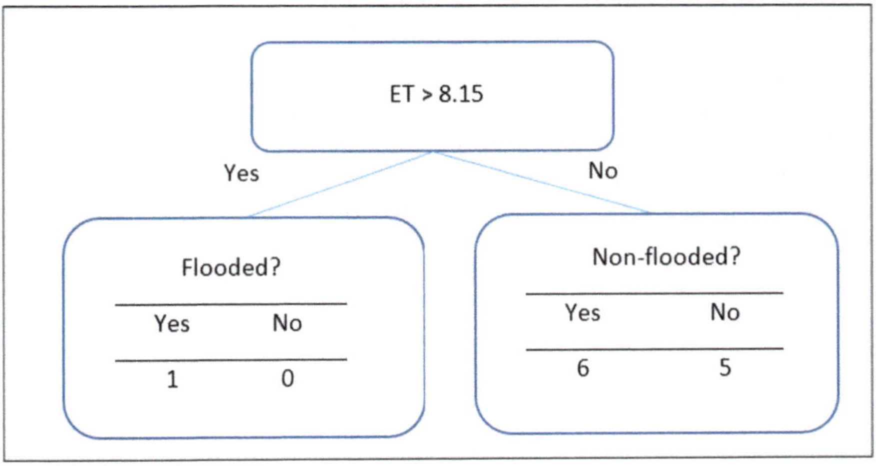

Fig. 4.10 Tree constructed using DNS at ET of 8.15

(fraction of negative examples ignored since there are no negative examples)

$$\text{Gini impurity}_{\text{right}} = \left(1 - \left(\frac{6}{11}\right)^2 - \left(\frac{5}{11}\right)^2\right) = 0.496$$

Thus, taking the weighted average:

$$\text{Gini impurity}_{\text{ET}=18.15} = \frac{0 \times 1 + 0.496 \times 11}{12} = 0.455$$

Table 4.7 ET in ascending order—Gini impurity

ET	y_i	Midpoint	Gini left	Gini right	Gini impurity
4.5	1				
5.4	1	4.95	0.496	0.000	0.455
5.5	1	5.45	0.500	0.000	0.417
7.8	1	6.65	0.494	0.000	0.370
8.5	−1	8.15	0.469	0.000	**0.313**
10.2	−1	9.35	0.490	0.320	0.419
11.2	1	10.7	0.500	0.444	0.472
12.4	−1	11.8	0.480	0.408	0.438
14.2	−1	13.3	0.500	0.469	0.479
14.5	1	14.35	0.444	0.494	0.481
16	−1	15.25	0.500	0.480	0.483
20.3	1	18.15	0.000	0.496	0.455

The best Gini impurity obtained when splitting the training examples using ET is 0.313 at a value of 8.15. Since DNS got a lower Gini impurity, it was decided to construct the first tree using DNS (DNS split at 4.7 has the lowest Gini impurity among the Gini impurities of both ET and DNS) (Fig. 4.11).

The total error (ε_t) is the sum of the weights of the incorrectly classified samples. All the samples here have the same weight of $\frac{1}{12}$.

It is understood that two of them are incorrectly classified since, on the left leaf, one training example is not a flooded point, and on the right leaf, one is a flooded point. The point DNS = 8.1 and ET = 14.2 fall to the left leaf (predicted flooded) but

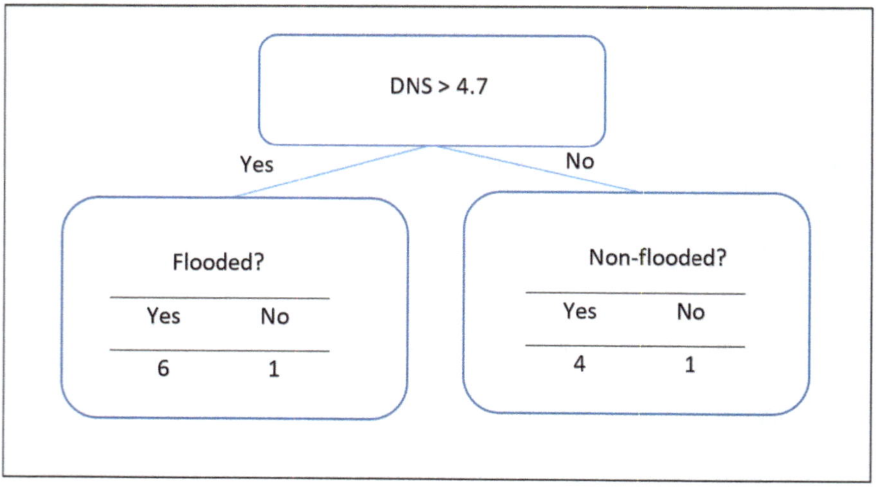

Fig. 4.11 The first tree constructed using DNS at DNS = 4.7

are not flooded. The point DNS = 4.3 and ET = 4.5 fall to the right leaf (predicted non-flooded) but are flooded. Other than these two points, the rest are correctly classified. Thus, the total error $\varepsilon_t = \frac{1}{12} + \frac{1}{12} = \frac{1}{6}$. Then, the weight of this DT (or stump) is calculated as (refer to Eq. 4.23)

$$\alpha_t = \frac{1}{2}Log\left(\frac{1-\varepsilon_t}{\varepsilon_t}\right) = \frac{1}{2}Log(5) = 0.805$$

The notation for the classification of a training sample based on a single tree is $h_t(x_i)$, where x_i is the ith training example at tth tree. The procedure is as follows:

(1) First, check the condition at the node (4.7 > 4.9), which is false; thus, move on to the leaf on the right side (corresponding to No). [4.7 is the value obtained to split the data since it had the least Gini impurity. Hence, comparing it with the DNS value of 4.9].

(2) The classification No refers to a class of -1; thus, the training example is classified as -1, which can be otherwise written as $h_1(x_6) = -1$. (x_6 refers to the point DNS = 4.9 and ET = 7.8). Here $h_1(x_6)$ refer to the first AdaBoost tree, and the subscript near x refers to the sixth training example.

Now, update the sample weights based on how the initial classification was made using Eq. 4.24, where $h_t(x_i)$ is the prediction by the tth DT of observation x_i. Equation 4.24 carries out the reduction of the sample weight of accurately classified samples and the increase of the sample weight of inaccurately classified samples. If substituted with $t = 1$, it refers to the first iteration of AdaBoost:

$$D_2(i) = \frac{D_1(i)e^{-\alpha_1 y_i h_1(x_i)}}{\sum_i D_1(i)e^{-\alpha_1 y_i h_1(x_i)}}$$

The new sample weights are calculated using Eq. 4.24. Since the denominator is the same for all the samples, it is first calculated independently. $D_1(i)$ is assumed to be $\frac{1}{12}$ for all samples since this is the first AdaBoost tree constructed. α_1 is already calculated as 0.805.

For the leaf on the left in Fig. 4.9 (or even Fig. 4.11), six points are flooded ($y_i = +1$) and classified as flooded ($h_1(x_i) = +1$), thus contributing to the term $6e^{(-0.805\times1\times1)}$. There is also one point that is not flooded ($y_i = -1$), but is classified as flooded ($h_1(x_i) = +1$), thus contributing to the term $e^{(-0.805\times-1\times1)}$. For the leaf on the right in Fig. 4.8 (or even Fig. 4.10), four points are not flooded ($y_i = -1$) and classified as non-flooded ($h_1(x_i) = -1$), thus contributing to the term $4e^{(-0.805\times-1\times-1)}$. There is also one point that is flooded ($y_i = +1$), but is classified as non-flooded ($h_1(x_i) = -1$), thus contributing to the term $e^{(-0.805\times1\times-1)}$.

The denominator $\sum_i D_t(i)e^{-\alpha_t y_i h_t(x_i)} = \frac{1}{12}[6e^{(-0.805\times1\times1)} + e^{(-0.805\times-1\times1)} + e^{(-0.805\times1\times-1)} + 4e^{(-0.805\times-1\times-1)}$.

The $\frac{1}{12}$ is the sample weight, $D_1(i)$ [first iteration and assumed weight], which is the same for all the samples at this stage. It is taken out as standard. In the first term,

$6e^{(-0.805\times1\times1)}$ refers to the six examples in the left leaf which have been classified correctly, with a weight of the DT (α_1) calculated as −0.805 using Eq. 4.23. The first one refers to y_1, the class, and the second 1 refers to $h_1(x_1)$. It is the class as predicted by the AdaBoost. Similarly, the other three terms are calculated. The above simplifies to

$$\frac{1}{12}[10\times0.447+2\times2.237]=0.745$$

After completion of the denominator (typical for all samples, i.e., 0.745), the numerator for each sample is calculated, and the new sample weight is updated accordingly (using Eq. 4.24). For instance, take the first example in Table 4.6. It has a DNS of 2.4, which, according to the constructed tree, is classified as No or −1, i.e., $h_1(x_1) = -1$. It is also the corresponding label −1, i.e., $y_i = -1$, which means it is a non-flooded location. In $h_1(x_1) = -1$, h_1 refers to the first AdaBoost tree, and the subscript of x_1 refers to the first training example. The −1 signifies that it has been predicted as non-flooded.

$$D_2(1)=D_2(2)=D_2(4)=D_2(5)=\frac{\frac{1}{12}e^{(-0.805\times-1\times-1)}}{0.745}=\frac{\frac{1}{12}\times0.447}{0.745}=0.05$$

The third example has a DNS value of 4.5, and according to our tree, it is classified as −1, i.e., $h_1(x_3) = -1$. However, the corresponding label is + 1, i.e., $y_i = +1$. It means that our tree incorrectly classified a positive training example as negative. This training example is incorrectly classified, so its sample weight is duly increased.

$$D_2(3)=D_2(8)=\frac{\frac{1}{12}e^{(-0.805\times1\times-1)}}{0.745}=0.25$$

Similarly, the rest of the sample weights are updated:

$$D_2(6)=D_2(7)=D_2(9)=D_2(11)=D_2(12)=\frac{\frac{1}{12}e^{(-0.805\times1\times1)}}{0.745}=0.05$$

$$D_2(10)=\frac{\frac{1}{12}e^{(-0.805\times-1\times1)}}{0.745}=\frac{\frac{1}{12}\times2.237}{0.745}=0.25$$

In addition, verify whether the newly updated weights add up to unity, which they do. The inference that can be made with the new weights is that correctly classified training examples have been given less weight (0.05). In comparison, the incorrectly classified training examples have been given a higher weight (25% chance). It ensures that more training examples must be correctly classified in the bootstrapped dataset.

A dataset of the identical size as the original is constructed. The training examples in the new dataset are chosen randomly according to their new sample weights. Repetition of a training sample in the new dataset is allowed. Create another dataset of the identical size as the previous dataset of size 12. This new dataset is created by randomly choosing from the original dataset. The sample weight is the probability that a new dataset is selected from the old dataset.

Whenever a new dataset is picked, two previously incorrectly classified training examples have a 25% chance of being selected. In comparison, the other training samples have a 5% chance. It is the first iteration of AdaBoost, and similarly multiple trees can be considered. The final prediction for each observation is predicted using Eq. 4.26:

$$H(x) = \text{sign}\left(\sum_{t=1}^{T} \alpha_t h_t(x)\right) \tag{4.26}$$

where H(x) represents the final prediction and $h_t(x)$ are the predictions by individual trees.

Use Eq. 4.26 to classify them if given an unseen test example. The below representation provides an idea of how unseen examples will be classified.

For example, if the point is DNS = 5 and ET = 5, calculate the probability based on the first tree constructed using Eq. 4.26.

$$H(x) = \text{sign}(0.805\ h_t(x)).$$

Traverse down the tree in Fig. 4.11. The DNS of our point is greater than 4.7. Thus, this tree classifies it as flooded, $h_t(x) = +1$.

Therefore $H(x) = \text{sign}(0.805 \times 1) = +1$.

It means that our AdaBoost classifier of just one tree would classify it as positive, i.e., flooded. As a note, the classification of AdaBoost after 50 trees was constructed, which is generated using code (Fig. 4.12).

4.8.2 *eXtreme Gradient Boosting*

XGBoost uses tree pruning to minimize the loss function using multiple weak learners (Osman et al., 2021; Wu et al., 2019). Examples are first classified such that identical residuals are in the same cluster and later then branched off (refer to Fig. 4.13). Detailed information about XGBoost is available from Madhuri et al. (2021), Mishra et al. (2024), Deb et al. (2024).

A similarity score is used as an attribute selection measure in this context and expressed (Eq. 4.27) as

$$\text{Similarity Score} = \frac{\left(\sum \text{Residual}_i\right)^2}{\sum\left[p_i' \times \left(1 - p_i'\right)\right] + \lambda_R} \tag{4.27}$$

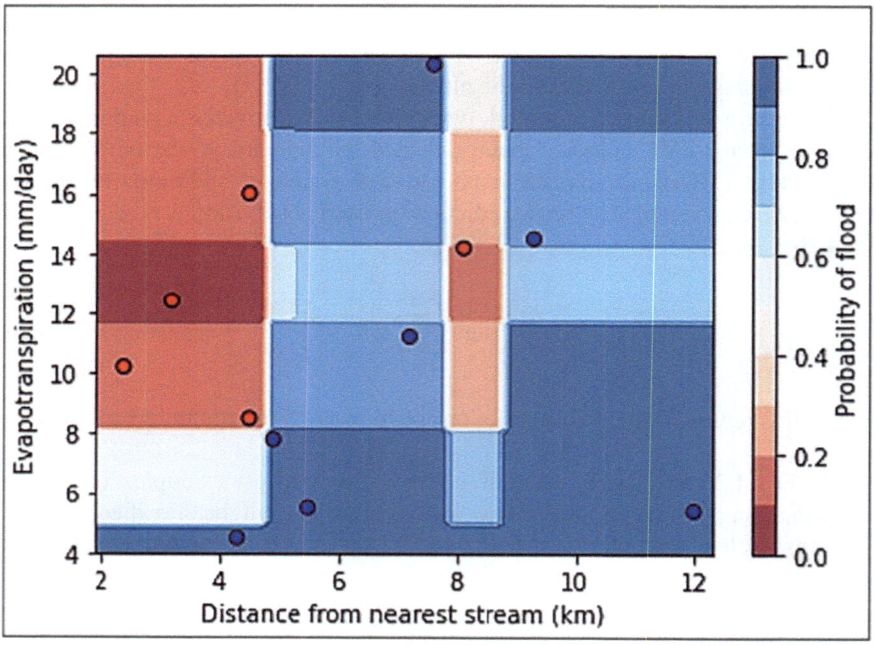

Fig. 4.12 Classification by AdaBoost after 50 trees were constructed (Red: Non-flooded; Blue: Flooded)

Here, λ_R is termed as the regularization parameter. p_i' is the prior probability estimated for the ith training example ($i = 1, 2, 3, ..n$) in that branch. Appropriate p_i' is assigned in the first iteration. The training examples are branched so that the information gain is the maximum at each branch till it can reach a maximum number of branches of the given tree. This gain is computed as Eq. (4.28).

$$\text{Gain} = \text{Similarity Score}_{\text{Left}} + \text{Similarity Score}_{\text{Right}} - \text{Similarity Score}_{\text{Root}} \quad (4.28)$$

Cover (min_child_weight or minimal number of residuals in each leaf) is computed as $\sum[p_i' \times (1 - p_i')]$. The leaf is taken out if the cover is less than the minimum, and the tree (or branch) is pruned if the Tree Complexity Parameter (γ) is higher than the gain at a branch and is one form of regularization to make the process more generalized. Larger λ_R lower the gain, thereby making pruning flexible. Lastly, output values (w) for all leaves are expressed as (Eq. (4.29))

$$w = \frac{\sum \text{Residual}_i}{\sum[p_i' \times (1 - p_i')] + \lambda_R} \quad (4.29)$$

The probabilities are fine-tuned in terms of Log of odds or Log odds (Eq. 4.30):

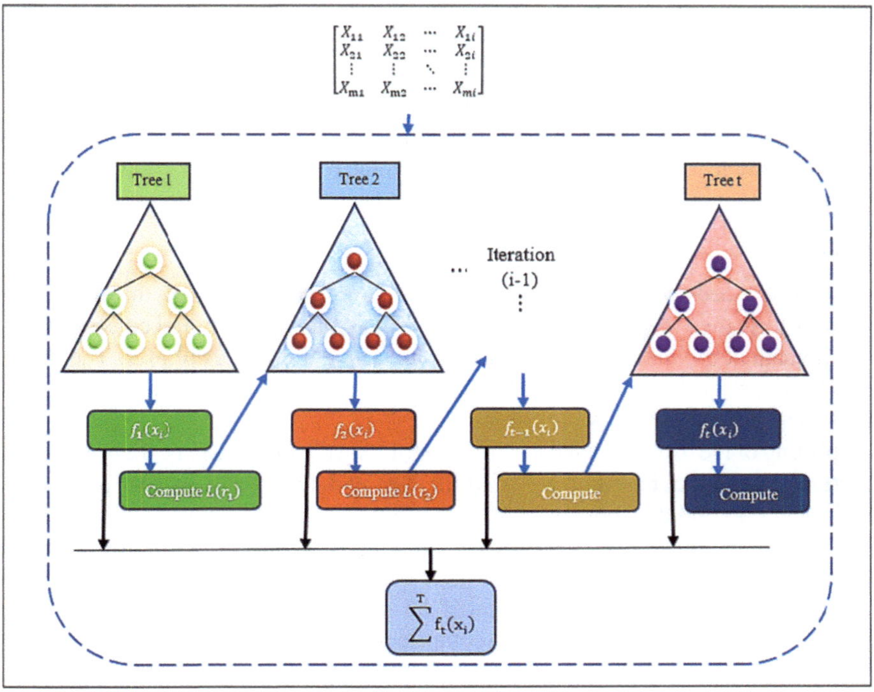

Fig. 4.13 Architecture of XGBoost (adapted from Deb et al., 2024 under CC BY 4.0 License)

$$\log\left(\frac{p}{1-p}\right) = \text{Log odds}. \quad (4.30)$$

The inverse of Eq. 4.30 is also used.

Another tree is added after each iteration. Accordingly, it is expressed as (Eq. 4.31)

$$\text{Log new odds} = \text{Log old odds} + \text{Learning rate}\left(\text{Output Value}_{\text{tree1}} + \text{Output Value}_{\text{tree2}} + \cdots + \text{Output Value}_{\text{treeN}}\right) \quad (4.31)$$

It can be observed from Eq. 4.31 that the output values from several trees are considered together in updating the Log odds. The mechanism of calculating Log odds and output values is demonstrated in the numerical example 4.6. Eventually, XGBoost targets to minimize the objective function (Eq. 4.32):

$$O(y_i, p_i, w) = \sum_{i=1}^{n} L(y_i, p_i) + \frac{1}{2}\lambda_R w^2 \quad (4.32)$$

where loss function, $L(y_i, p_i)$, is expressed in Eq. 4.33:

$$L(y_i, p_i) = [-y_i \log(p_i) + (1 - y_i)\text{Log}(1 - p_i)] \tag{4.33}$$

Numerical Problem 4.6. Classify the datasets presented in Table 4.8 using XGBoost. Features considered are DNS and ET. The initial probability prediction for all values is 0.5, and the learning rate is 0.3. Employ similarity score as an attribute selection measure.

Solution:

The feature needs to be identified to initiate the construction of the tree. Like AdaBoost, arrange the values for each feature in ascending order and consider splitting the data at the mean of each pair of consecutive training examples. The initial probability prediction for all values is 0.5, on which the residuals can be calculated (Table 4.9).

Splitting of the data is needed to start constructing the tree. It can be done at several points. For example, based on ET, split the data at 8.15, which has four points less than it and eight points above it (8.15 is the mean of the consecutive ET values 7.8 and 8.5). The middle point is constructed using these mean values (Table 4.10). The gain is calculated at each point where the data is split (based on the middle point). For example, split the tree at 9.35 (mean of 8.5 and 10.2). Before calculating the gain, the similarity scores of the root node (comprising of the residuals of all the data, left leaf and right leaf) can be computed. λ_R is assumed to be zero for demonstration purposes. The root node consists of all the points, whereas the left leaf consists of the points less than the threshold chosen, and the right leaf consists of the points more significant than the threshold.

Table 4.8 Datasets for the numerical problem

Dataset	DNS	ET	Did flood occur?	Observed y_i
1	2.4	10.2	No	0
2	12	5.4	Yes	1
3	4.5	16	No	0
4	7.6	20.3	Yes	1
5	9.3	14.5	Yes	1
6	4.9	7.8	Yes	1
7	8.1	14.2	No	0
8	4.3	4.5	Yes	1
9	3.2	12.4	No	0
10	5.5	5.5	Yes	1
11	7.2	11.2	Yes	1
12	4.5	8.5	No	0

Table 4.9 Residual probabilities and middle points of ET

ET	y_i	p_i	Residual y_i -p_i	Middle point
4.5	1	0.5	0.5	–
5.4	1	0.5	0.5	4.95
5.5	1	0.5	0.5	5.45
7.8	1	0.5	0.5	6.65
8.5	0	0.5	−0.5	8.15
10.2	0	0.5	−0.5	9.35
11.2	1	0.5	0.5	10.7
12.4	0	0.5	−0.5	11.8
14.2	0	0.5	−0.5	13.3
14.5	1	0.5	0.5	14.35
16	0	0.5	−0.5	15.25
20.3	1	0.5	0.5	18.15

$$\text{Similarity score}_{\text{Root}} = \frac{\left(\sum \text{Residual}_i\right)^2}{\sum\left[\text{p}_i' \times \left(1 - \text{p}_i'\right)\right] + \lambda_{\text{R}}} = \frac{(0.5 + 0.5 \ldots - 0.5 + 0.5)^2}{12(0.5 \times 0.5) + 0} = 0.333$$

$$\text{Similarity score}_{\text{Left}} = \frac{\left(\sum \text{Residual}_i\right)^2}{\sum\left[\text{p}_i' \times \left(1 - \text{p}_i'\right)\right] + \lambda_{\text{R}}} = \frac{(0.5 + 0.5 + 0.5 + 0.5 - 0.5)^2}{5(0.5 \times 0.5) + 0} = 1.8$$

$$\text{Similarity score}_{\text{Right}} = \frac{\left(\sum \text{Residual}_i\right)^2}{\sum\left[\text{p}_i' \times \left(1 - \text{p}_i'\right)\right] + \lambda_{\text{R}}} = \frac{(-0.5 + 0.5 - 0.5 - 0.5 + 0.5 - 0.5 + 0.5)^2}{7(0.5 \times 0.5) + 0} = 0.143$$

After calculating the similarity scores, the gain can be calculated using Eq. 4.28.

$$\text{Gain} = \text{Similarity Score}_{\text{Left}} + \text{Similarity Score}_{\text{Right}} - \text{Similarity Score}_{\text{Root}} = 1.8 + 0.143 - 0.334 = 1.609$$

Similarly, the gain is calculated for both features at every point, as shown in Tables 4.10 and 4.11.

Table 4.10 Gain values of the first branch for ET

ET	y_i	Middle	Left score	Right score	Root	Gain
4.5	1	–	–	–	–	–
5.4	1	4.95	1.000	0.091	0.333	0.758
5.5	1	5.45	2.000	0.000	0.333	1.667
7.8	1	6.65	3.000	0.111	0.333	2.778
8.5	0	8.15	4.000	0.500	0.333	4.167 (Highest gain)
10.2	0	9.35	1.800	0.143	0.333	1.609
11.2	1	10.7	0.667	0.000	0.333	0.333
12.4	0	11.8	1.286	0.200	0.333	1.152
14.2	0	13.3	0.500	0.000	0.333	0.167
14.5	1	14.35	0.111	0.333	0.333	0.111
16	0	15.25	0.400	0.000	0.333	0.067
20.3	1	18.15	0.091	1.000	0.333	0.758

Table 4.11 Gain values of first branch for DNS

DNS	y_i	Middle	Left score	Right score	Root	Gain
2.4	0	–	–	–	–	–
3.2	0	2.8	1.000	0.818	0.333	1.485
4.3	1	3.75	2.000	1.600	0.333	3.267
4.5	0	4.4	0.333	1.000	0.333	1.000
4.5	0	4.5	0.333	1.000	0.333	1.000
4.9	1	4.7	1.800	3.571	0.333	5.038 (Highest gain)
5.5	1	5.2	0.667	2.667	0.333	3.000
7.2	1	6.35	0.143	1.800	0.333	1.610
7.6	1	7.4	0.000	1.000	0.333	0.667
8.1	0	7.85	0.111	0.333	0.333	0.111
9.3	1	8.7	0.000	2.000	0.333	1.667
12	1	10.65	0.091	1.000	0.333	0.758

The first tree is built using the split, which gives the highest gain at a DNS value of 4.7, with a gain of 5.038, and the residuals are split, as per Fig. 4.14.

The leaves created can be further divided (Table 4.12), similar to how they were separated earlier, based on a split that gives a maximum gain.

First, split the right leaf (arbitrarily). The points corresponding to the right leaf are given in Tables 4.13–4.16. The split at which the highest gain is obtained for these points is found by first calculating the gain at every possible split.

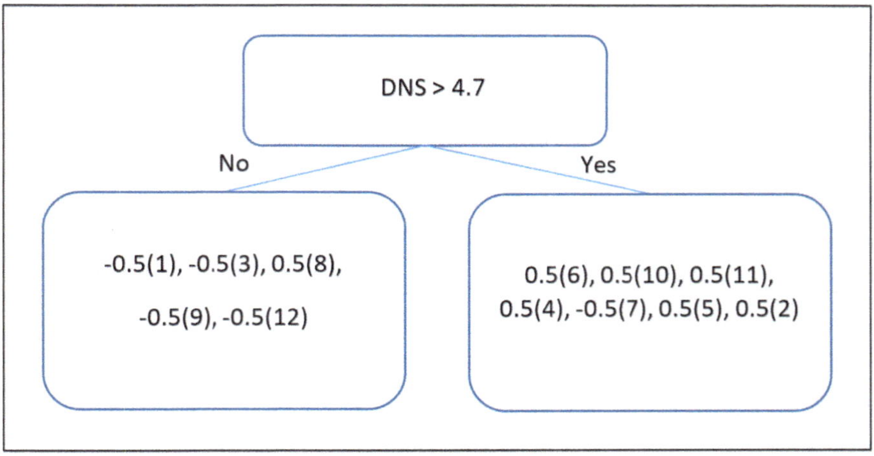

Fig. 4.14 First branching of the tree (0.5 and −0.5 are the residuals; brackets denote the dataset number)

Table 4.12 Points belonging to the left leaf of the first branch

DNS	ET	y_i
2.4	10.2	0
4.5	16.0	0
4.3	4.5	1
3.2	12.4	0
4.5	8.5	0

Table 4.13 Gain values of the second branch (ET)

ET	y_i	Middle value	Left score	Right score	Root	Gain
4.5	1	–	–	–	–	–
8.5	0	6.5	1.00	4.00	1.80	3.2
10.2	0	9.35	0.00	3.00	1.80	1.2
12.4	0	11.3	0.33	2.00	1.80	0.533
16	0	14.2	1.00	1.00	1.80	0.2

Table 4.14 Gain values of the second branch (DNS)

DNS	y_i	Middle value	Left score	Right score	Root	Gain
2.4	0	–	–	–	–	–
3.2	0	2.8	1.000	1.000	1.800	0.2
4.3	1	3.75	2.000	0.333	1.800	0.533
4.5	0	4.4	0.333	2.000	1.800	0.533
4.5	0	4.5	0.333	2.000	1.800	0.533

Table 4.15 Gain values of the third branch (ET)

ET	y_i	Middle value	Left score	Right score	Root	Gain
5.4	1	–	–	–	–	–
5.5	1	5.45	1.000	2.667	3.571	0.095
7.8	1	6.65	2.000	1.800	3.571	0.229
11.2	1	9.5	3.000	1.000	3.571	0.429
14.2	0	12.7	4.000	0.333	3.571	0.762
14.5	1	14.35	1.800	2.000	3.571	0.229
20.3	1	17.4	2.667	1.000	3.571	0.095

Table 4.16 Gain values of the third branch (DNS)

DNS	y_i	Middle value	Left score	Right score	Root	Gain
4.9	1	–	–	–	–	–
5.5	1	5.2	1.000	2.667	3.571	0.095
7.2	1	6.35	2.000	1.800	3.571	0.229
7.6	1	7.4	3.000	1.000	3.571	0.429
8.1	0	7.85	4.000	0.333	3.571	0.762
9.3	1	8.7	1.800	2.000	3.571	0.229
12	1	10.65	2.667	1.000	3.571	0.095

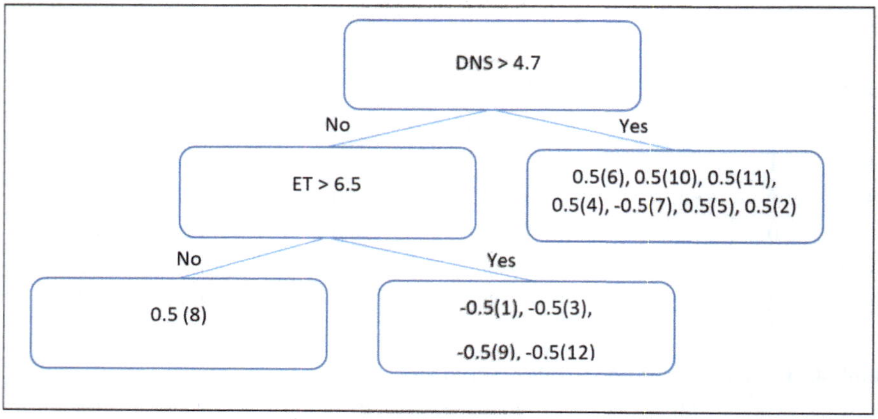

Fig. 4.15 Second branching of the tree

The split with the highest gain is made at an ET value of 6.5 (middle value), with a gain of 3.2. It is observed that similar residuals end up in the same leaves (Fig. 4.15).

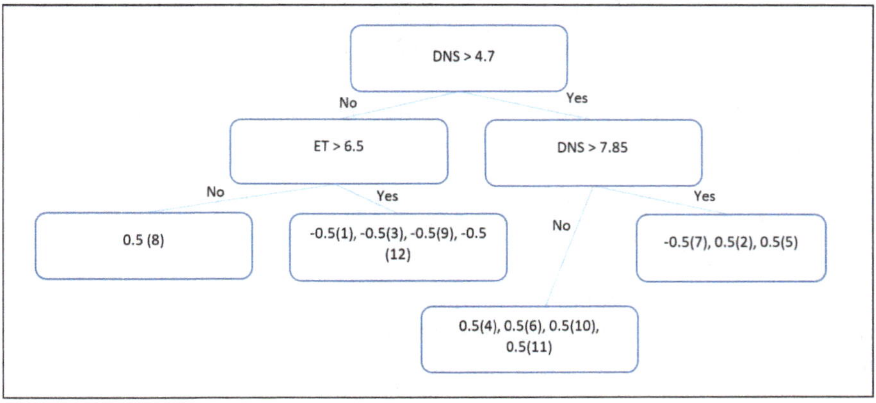

Fig. 4.16 Complete branching of the tree

The tree may also be expanded on the right leaf using the same procedure. Gain at each split is calculated as follows:

It so happened that the one negative class (0) point (DNS: 8.7, ET: 14.5; dataset number 7) in this leaf occurs in the same place in ascending order for both features. It means that all the scores for both features will be the same. Thus, the gain calculated for both features is the same. Accordingly, DNS is arbitrarily picked. The split is made at a DNS value of 7.85, with a gain of 0.762 (Fig. 4.16).

First, after going through the root node (which contained all the training examples), it is decided to split at DNS > 4.7 (at level 1). At this level, there are two leaves. The leaf on the left contains five training examples, and the leaf on the right contains 7. It is found that the criterion by which both leaves could be split [ET > 6.5] and [DNS > 7.85]. After splitting both leaves one more time, the status is level 2. At this level, there are four leaves. It can be labelled as 1, 2, 3, and 4 from left to right. The branching stops when the maximum tree depth parameter is reached (here, it is 2).

Each leaf in a total of four leaves is used to calculate its output value using Eq. 4.29. The output value of the first leaf is (dataset number: 8; residual is 0.5)

$$w = \frac{\sum \text{Residual}_i}{\sum[p_i' \times (1 - p_i')] + \lambda_R} = \frac{0.5}{0.25} = 2$$

The output value of the second leaf is calculated as (dataset number: 1, 3, 9, 12; all residuals are −0.5)

$$w = \frac{\sum \text{Residual}_i}{\sum[p_i' \times (1 - p_i')] + \lambda_R} = \frac{-0.5 - 0.5 - 0.5 - 0.5}{0.25 + 0.25 + 0.25 + 0.25} = -2$$

The output value of the third leaf is (dataset number: 4, 6, 10, 11; all residuals are 0.5)

$$w = \frac{\sum \text{Residual}_i}{\sum[p_i' \times (1 - p_i')] + \lambda_R} = \frac{0.5 + 0.5 + 0.5 + 0.5}{0.25 + 0.25 + 0.25 + 0.25} = 2$$

The output value of the fourth leaf is (dataset number 2,5,7; residuals are −0.5, 0.5, and 0.5)

$$w = \frac{\sum \text{Residual}_i}{\sum[p_i' \times (1 - p_i')] + \lambda_R} = \frac{-0.5 + 0.5 + 0.5}{0.25 + 0.25 + 0.25} = 0.667$$

Now that the output values of each last leaf have been made, an update on the estimated probability can be made. This process is done in terms of Log odds of the probabilities. Equation 4.30 shows how Log odds are calculated.

$$\log\left(\frac{p}{1 - p}\right) = \text{Log odds}$$

Since the initial probability estimate of all the training points is 0.5, the Log odds of all the training points are $\log(\frac{0.5}{0.5}) = 0$. The Log newodds are found by updating them with the output values and learning rate of 0.3 (Eq. 4.31).

$$\text{Log newodds} = \text{Log oldodds} + \text{Learning rate} \times \text{Output Value}$$

Develop knowledge (Table 4.17) using the tree created (refer to Fig. 4.16) and assign an output value to each training example. Then, using the output value, the Log odds are calculated. Furthermore, these are converted back into the form of probability to receive the new estimates of the training examples using the following:

$$\text{Probability} = \frac{e^{\text{Log newodds}}}{1 + e^{\text{Log newodds}}} \tag{4.34}$$

Table 4.17 presents the updated probabilities. Figure 4.17 illustrates the results after constructing 100 trees. It is noted that the newly predicted values are nearer to the actual class of each training example.

The procedure mentioned above is for the first tree. Other trees are also constructed by calculating new residuals using the newly estimated probability values.

If the class of an unknown point is to be tested, it is run through all trees constructed, and using the output values given by each tree, the probability of the new point is found. The below representation provides an idea of how XGBoost will classify unseen examples.

For example, if it was given the training point with DNS and ET values as 5.0, the process of how its value can be predicted using the first tree calculated can be visualized (Fig. 4.16). DNS > 4.7 computes to true, so move to the right leaf of the first level. DNS > 7.85 [in Table 4.16, the middle value of 7.6 and 8.1 is 7.85; hence it is chosen since it has the highest gain] computes to false, so move to the second level's left leaf (labelled 3). The output value of the third leaf has been calculated as

Table 4.17 Updated probabilities

Dataset	DNS	ET	y_i	p_i	*Log old odds*	Leaves	Output value	*Log new odds*	Updated probability p_i
1	2.4	10.2	0	0.5	0	2	−2	−0.6	0.354
2	12	5.4	1	0.5	0	4	0.667	0.20	0.550
3	4.5	16	0	0.5	0	2	−2	−0.6	0.354
4	7.6	20.3	1	0.5	0	3	2	0.6	0.646
5	9.3	14.5	1	0.5	0	4	−2	−0.6	0.354
6	4.9	7.8	1	0.5	0	3	2	0.6	0.646
7	8.1	14.2	0	0.5	0	4	0.667	0.20	0.550
8	4.3	4.5	1	0.5	0	1	−2	−0.6	0.354
9	3.2	12.4	0	0.5	0	2	−2	−0.6	0.354
10	5.5	5.5	1	0.5	0	3	2	0.6	0.646
11	7.2	11.2	1	0.5	0	3	2	0.6	0.646
12	4.5	8.5	0	0.5	0	2	−2	−0.6	0.354

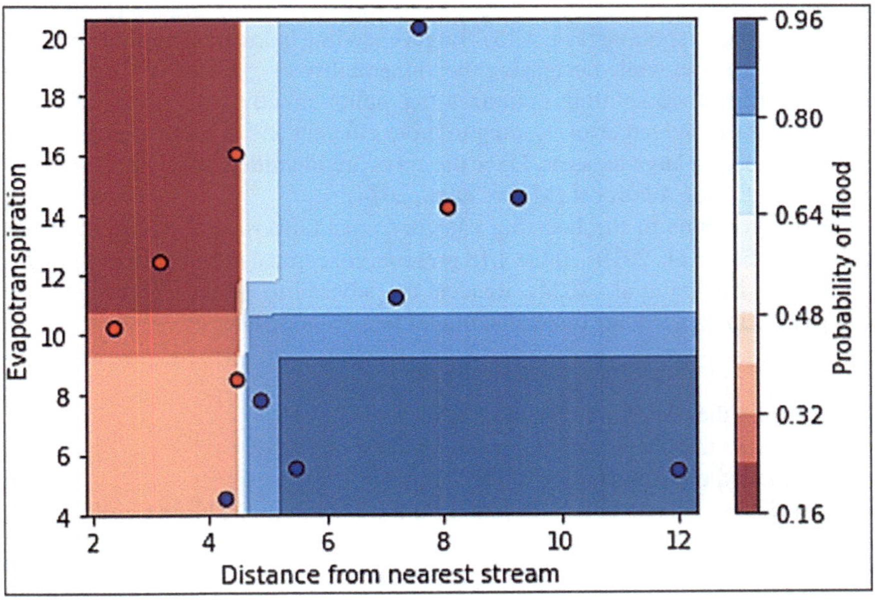

Fig. 4.17 XGBoost—results after construction of 100 trees (Red: Non-flooded; Blue: Flooded)

two, as shown above. Then, use Eq. 4.31 to update the probability from 0.5 to a new value.

$$\text{Log newodds} = \text{Log oldodds} + \text{Learning rate} \times \text{Output Value}$$

Expanding each term here:

$$\text{Log newodds} = \log\left(\frac{0.5}{1-0.5}\right) + 0.3 \times 2 = 0.6$$

Then calculate the updated probability from the Log newodds Eq. (4.34)

$$\text{Probability} = \frac{e^{\text{Log newodds}}}{1 + e^{\text{Log newodds}}} = \frac{e^{0.6}}{1 + e^{0.6}} = 0.645$$

Thus, based on the first tree, the points DNS = 5 and ET = 5 are estimated to have a 0.645 chance of flood.

4.8.3 *Categorical Boosting*

CatBoost (Mehraein et al., 2022), a gradient boosting algorithm, is constructed on the iterative framework of AdaBoost with the incorporation of (a) target encoding and (b) gradient-based splitting (Fig. 4.18). Target encoding helps enhance the predictive ability of the model while decreasing the vulnerability to over-fitting. Complimentarily, gradient-based splitting optimizes tree splits directly with reference to the gradient of the weighted error, leading to more efficient and accurate tree construction, especially on large datasets. Once the splits are identified, the iterative process is similar to that of AdaBoost (Mishra et al., 2024).

Other algorithms in the boosting category are LGBoost and NGBoost (Duan et al., 2020; Fan et al., 2019). Table 4.18 presents conceptual differences in Boosting Algorithms (Mishra et al., 2024). Readers are advised to go through the relevant sources to gain an adequate understanding of these algorithms.

Revision Questions and Exercise Problems

4.1 Discuss the philosophy behind CNN.
4.2 What are the filters that are commonly used in CNN?
4.3 Explain the architecture of CNN.
4.4 Explain the mathematical equation for the convolutional layer for ith data.
4.5 What is the pooling layer? What is average and maximum pooling?
4.6 What is the Softmax function layer? What is its use?
4.7 What is loss function? How is it going to help the efficacy of the ML algorithm?
4.8 What are the parameters governing CNN? According to you, which parameter has a significant effect on the efficacy of the algorithm?
4.9 What are the purposes of the learning and dropout rates? Are they related? If so, how?
4.10 Employ the CNN to establish a relationship between the input (traffic flow, lane length) and output (congestion level). Use the maximum pooling method.

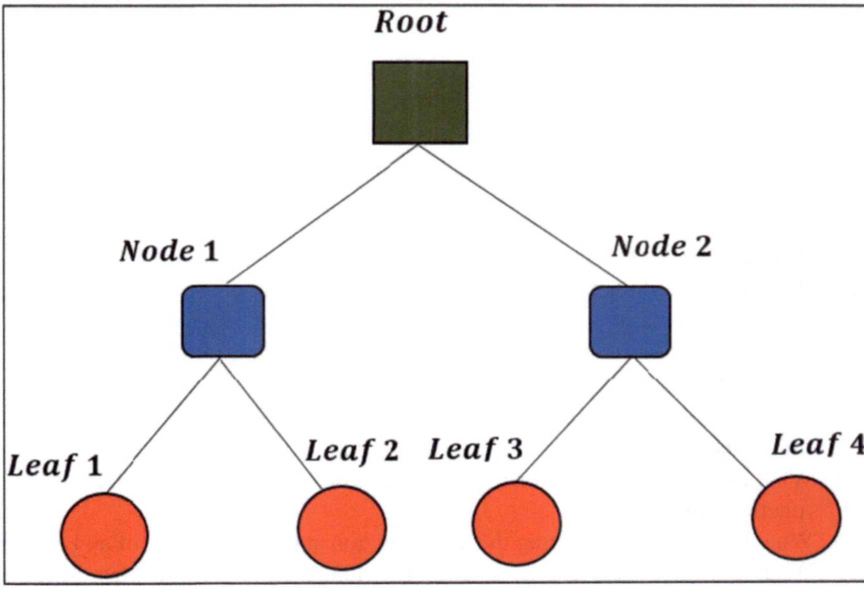

Fig. 4.18 Tree formation in CatBoost (adapted from Mishra et al., 2024 under CC BY-NC-ND 4.0 License)

Table 4.18 Conceptual differences in boosting algorithms (adapted from Mishra et al., 2024 under CC BY-NC-ND 4.0 License)

Features	Algorithms				
	AdaBoost	CatBoost	LGBoost	NGBoost	XGBoost
Formation of decision tree	Asymmetric level-wise	Symmetric level-wise growth	Asymmetric leaf-wise growth	Level-wise growth	Depth-wise growth
Splitting metliod	Greedy splitting method	Greedy splitting method	Gradient-based One-Side sampling	Natural gradient	Histogram based
Handling categorical features	No	Yes	No	No	No
Regularization	No	Yes	Yes	Yes	Yes
Memory consumption	Low	High	Low	High	Moderate
Feature importance	Available	Available	Available	Available	Available
Scalability	Fast	Moderate	Fast	Low	Fast

x_1	x_2	D
8	4	0.75
12	8	1.00

4.11 What is RNN? Explain briefly the mathematics behind the same.
4.12 What is the disadvantage of RNN that prompted the usage of LSTM?
4.13 What is the architecture of LSTM? Explain the mathematical philosophy in detail.
4.14 How can LSTM exploration be utilized to capture long-term dependencies?
4.15 What is the meaning of signal in LSTM?
4.16 What are input, forget, and output gates? How do they differ in their functionality?
4.17 What are final, new memory, and hidden cells? How do they differ in their functionality?
4.18 What is the purpose of using the *tanh* function at the output gate of an LSTM?
4.19 What are the parameters of LSTM? Mention them with their specific purpose. Which parameter significantly affects the outcome? Similarly, which parameter has the most negligible influence on the outcome?
4.20 What is the physical meaning of weights in LSTM?
4.21 The problem is related to eye movement recognition. Here, the input is skin temperature, x_t = (5,6,7), and output is eye movement, y_t = (2,3,4). Apply LSTM with three hidden units, λ_{t-1} = (1,2,3) and μ_{t-1} = (5,5,5). Assume the weights and parameters suitably.
4.22 What is the mathematical basis of GRU? Is it simpler than LSTM? If so, discuss in what aspects?
4.23 What is the meaning of hybridization in ML algorithms? Do you think it will improve performance compared to the individual algorithms?
4.24 What is the meaning of boosting in the ML framework?
4.25 Mention the names of three boosting algorithms. Compare them with three salient features.
4.26 What is the meaning of correctly and incorrectly classified samples in boosting algorithms? How are they going to affect the quality of output?
4.27 What is the meaning of a bootstrapped sampling?
4.28 What is Gini impurity? Can it have a value greater than 1?
4.29 A leaf has ten training examples that belong to the positive class (1) and six training examples that belong to the negative class (−1). Assume suitable data, if any compute Gini Index.
4.30 Is lower Gini impurity preferred or higher? Why?
4.31 What is the purpose of middle values in boosting-related problems? On what basis can the weighted average of the Gini impurities be computed?

4.32 Mention one distinct advantage and disadvantage in AdaBoost that affects the accuracy of the outcome. What possible extensions can be explored to make the algorithm more robust?
4.33 What are the residuals in XGBoost?
4.34 What is a similarity score and regularization parameter?
4.35 What is information gain, and how is it related to similarity scores?
4.36 What is the Tree Complexity Parameter?
4.37 What is the *Logodds* ? Is it related to probability? If yes, how is it related? If the *Logodds* is 0.8, what is the probability?
4.38 What are the objective functions of boosting algorithms?
4.39 Mention one distinct advantage and disadvantage in XGBoost that affects the accuracy of the outcome.
4.40 What possible extensions can be explored to make the boosting algorithms more robust?
4.41 The problem is related to the stability of the foundation. Two features are considered: effective vertical stresses and earthquake magnitude. Seven different locations are the datasets. Classify the datasets using AdaBoost and XGBoost (refer to Table 4.19).

Advanced Review Questions

4.42 Can you explore pooling methods other than average and maximum pooling? If yes, how are they better than average pooling?
4.43 What is the physical significance of the height and width of filters in CNN?
4.44 What is meant by dense vector? What is the physical meaning of weights and biases of dense vectors?
4.45 What is the purpose of termination criteria? Mention different types that can be explored.
4.46 How do epochs impact the learning and dropout rates? Can any relationship be established in this regard?
4.47 Is hyperparameter tuning necessary for improving the efficacy of algorithms? Do you think the computational burden will increase with hyperparameter tuning?

Table 4.19 Information about features and foundation status

Dataset	Effective vertical stresses	Earthquake magnitude	y_i	Is foundation safe?
1	2.8	10.8	1	Yes
2	12.6	5.8	0	No
3	4.8	16.8	1	Yes
4	7.8	20.8	0	No
5	9.9	14.8	1	Yes
6	6.9	8.8	0	No
7	8.8	16.2	0	No

4.48 Is the hybridization of CNN and LSTM the same as that of LSTM and CNN? Can you elaborate on the same? In that case, how may architecture be changed?
4.49 How to determine optimum values in the case of support vector regression?
4.50 Can you suggest further hybridizations between ML algorithms (besides those mentioned in this chapter)? If so, cite the basis of the same and the logical advantages of the same over the standalone algorithms.
4.51 What is meant by bias? Does boosting algorithms reduce bias?
4.52 Is using the learning rate in gradient boosting essential to get an optimum output? Elaboration is suggested.
4.53 Can you develop a boosting algorithm that simultaneously minimizes computational complexity and facilitates accurate predictions?

References

Aldrees, A., Awan, H. H., Javed, M. F., & Mohamed, A. M. (2022). Prediction of water quality indexes with ensemble learners: Bagging and boosting. *Process Safety and Environmental Protection, 168*, 344–361.

Alzubaidi, L., Zhang, J., Humaidi, A. J., Al-Dujaili, A., Duan, Y., Al-Shamma, O., Santamaría, J., Fadhel, M. A., Al-Amidie, M., & Farhan, L. (2021). Review of deep learning: Concepts, CNN architectures, challenges, applications, future directions. *Journal of Big Data, 8*, 53.

Deb, D., Vasan, A., & Raju, K. S. (2024). Daily reservoir inflow prediction using stacking ensemble of machine learning algorithms. *Journal of Hydroinformatics*, 26, 972–997 [open access paper under CC BY 4.0 License]

Ding, Y., Zhu, H., Chen, R., & Li, R. (2022). An efficient adaBoost algorithm with the multiple thresholds classification. *Applied Sciences, 12*, 5872.

Duan, T., Anand, A., Ding, D. Y., Thai, K. K., Basu, S., Ng, A., & Schuler, A. (2020). Ngboost: Natural gradient boosting for probabilistic prediction. Proceedings of the 37th *International Conference on Machine Learning, July 13–18, 2020 (Virtual Event)* 119, 2690–2700.

Fan, J., Ma, X., Wu, L., Zhang, F., Yu, X., & Zeng, W. (2019). Light gradient boosting machine: An efficient soft computing model for estimating daily reference evapotranspiration with local and external meteorological data. *Agricultural Water Management, 225*, 105758.

Horchreiter, S., & Schmidhuber, J. (1997). Long short-term memory. *Neural Computation, 9*, 1735–1780.

Islam, M., Nooruddin, S., Karray, F., & Muhammad, G. (2022). Human activity recognition using tools of convolutional neural networks: A state of the art review, data sets, challenges, and future prospects. *Computers in Biology and Medicine, 149*, 106060.

Madhuri, R., Sistla, S., & Raju, K. S. (2021). Application of machine learning algorithms for flood susceptibility assessment and risk management. *Journal of Water and Climate Change, 12*, 2608–2623.

Mehraein, M., Mohanavelu, A., Naganna, S. R., Kulls, C., & Kisi, O. (2022). Monthly streamflow prediction by metaheuristic regression approaches considering satellite precipitation data. *Water, 14*, 3636.

Mishra, B. R., Vogeti, R. K., Jauhari, R., Raju, K. S., & Kumar, D. N. (2024). Boosting algorithms for projecting streamflow in the lower godavari basin for different climate change scenarios. *Water Science and Technology*, 89, 613–634 [open access paper under CC BY-NC-ND 4.0 License]

Neu, D. A., Lahann, J., & Fettke, P. (2022). A systematic literature review on state-of-the-art deep learning methods for process prediction. *Artificial Intelligence Review, 55*, 801–827.

Orojo, O., Tepper, J., McGinnity, T. M., & Mahmud, M. (2023). The multi-recurrent neural network for state-of-the-art time-series processing. *Procedia Computer Science, 222*, 488–498.

Osman, A. I. A., Ahmed, A. N., Chow, M. F., Huang, Y. F., & El-Shafie, A. (2021). Extreme gradient boosting (Xgboost) model to predict the groundwater levels in Selangor Malaysia. *Ain Shams Engineering Journal, 12*, 1545–1556.

Ren, Y., Zeng, S., Liu, J., Tang, Z., Hua, X., Li, Z., & Xia, J. (2022). Mid-to long-term runoff prediction based on deep learning at different time scales in the upper Yangtze River Basin. *Water, 14*, 1692.

Roy, D. K., Sarkar, T. K., Kamar, S. S. A., Goswami, T., Muktadir, M. A., Al-Ghobari, H. M., Alataway, A., Dewidar, A. Z., El-Shafei, A. A., & Mattar, M. A. (2022). Daily prediction and multi-step forward forecasting of reference evapotranspiration using LSTM and Bi-LSTM algorithms. *Agronomy, 12*, 594.

Shen, G., Tan, Q., Zhang, H., Zeng, P., & Xu, J. (2018). Deep Learning with Gated Recurrent Unit Networks for Financial Sequence Predictions. *Procedia Computer Science, 131*, 895–903.

Van Houdt, G., Mosquera, C., & Nápoles, G. (2020). A review on the long short-term memory model. *Artificial Intelligence Review, 53*, 5929–5955.

Van, S. P., Le, H. M., Thanh, D. V., Dang, T. D., Loc, H. H., & Anh, D. T. (2020). Deep learning convolutional neural network in rainfall-runoff modelling. *Journal of Hydroinformatics, 22*, 541–561.

Vogeti, R. K., Jauhari, R., Mishra, B. R., Raju, K. S., & Kumar, D. N. (2024). Deep learning algorithms and their fuzzy extensions for streamflow prediction in climate change framework. *Journal of Water and Climate Change*, 15, 832–848 [open access paper under CC BY-NC-ND 4.0 License].

Wu, L., Peng, Y., Fan, J., & Wang, Y. (2019). Machine learning algorithms for the estimation of monthly mean daily reference evapotranspiration based on cross-station and synthetic data. *Hydrology Research, 50*, 1730–1750.

Yu, Y., Si, X., Hu, C., & Zhang, J. (2019). A review of recurrent neural networks: LSTM cells and network architectures. *Neural Computation, 31*, 1235–1270.

Suggested Further Reading

Ahmed, S. F., Alam, M. S. B., Hassan, M., Rozbu, M. R., Ishtiak, T., Rafa, N., Mofijur, M., Ali, A. B. M. S., & Gandomi, A. H. (2023). Deep learning modelling techniques: Current progress, applications, advantages, and challenges. *Artificial Intelligence Review, 56*, 13521–13617.

Alom, M. Z., Taha, T. M., Yakopcic, C., Westberg, S., Sidike, P., Nasrin, M. S., Hasan, M., Van Essen, B. C., Awwal, A. A. S., & Asari, V. K. (2019). A state-of-the-art survey on deep learning theory and architectures. *Electronics, 8*, 292.

Alshehri, A. S., & You, F. (2021). Paradigm shift: The promise of deep learning in molecular systems engineering and design. *Frontiers in Chemical Engineering*, 3–2021.

Chen, L., Li, S., Bai, Q., Yang, J., Jiang, S., & Miao, Y. (2021). Review of image classification algorithms based on convolutional neural networks. *Remote Sensing, 13*, 4712.

Islam, M. A., Rashid, S. I., Hossain, N. U. I., Fleming, R., & Sokolov, A. (2023). An integrated convolutional neural network and sorting algorithm for image classification for efficient flood disaster management. *Decision Analytics Journal, 7*, 100225.

Klaiber, J., & Dinther, C. V. (2023). Deep learning for variable renewable energy: A systematic review. *ACM Computing Surveys, 56*, 1–37.

Kumar, V., Azamathulla, H. M., Sharma, K. V., Mehta, D. J., & Maharaj, K. T. (2023). The state of the art in deep learning applications, challenges, and future prospects: A comprehensive review of flood forecasting and management. *Sustainability, 15*, 10543.

Naranjo-Torres, J., Mora, M., Hernández-García, R., Barrientos, R. J., Fredes, C., & Valenzuela, A. (2020). A review of convolutional neural network applied to fruit image processing. *Applied Sciences, 10*, 3443.

Nasreen, G., Haneef, K., Tamoor, M., & Irshad, A. (2023). Review: A Comparative study of state-of-the-art skin image segmentation techniques with CNN. *Multimedia Tools and Applications, 82*, 10921–10942.

Olorunnimbe, K., & Viktor, H. (2023). Deep learning in the stock market—a systematic survey of practice. *Backtesting, and Applications, Artificial Intelligence Review, 56*, 2057–2109.

Salau, L., Hamada, M., Prasad, R., Hassan, M., Mahendran, A., & Watanobe, Y. (2022). State-of-the-art survey on deep learning-based recommender systems for E-learning. *Applied Sciences, 12*, 11996.

Thakur, R. S., Yadav, R. N., & Gupta, L. (2019). State-of-art analysis of image denoising methods using convolutional neural networks. *IET Image Processing, 13*, 2367–2380.

Valizadeh, M., & Wolff, S. J. (2022). Convolutional Neural Network applications in additive manufacturing: A review. *Advances in Industrial and Manufacturing Engineering, 4*, 100072.

Chapter 5
Fuzzy-Based Modelling Algorithms

5.1 Introduction

The chapter provides an insight into fuzzy logic-based approaches, namely, Fuzzification and Defuzzification, Adaptive Neuro-Fuzzy Inference System (ANFIS), Fuzzy Cognitive Mapping (FCM), Optimization, and its fuzzy extension. This chapter also briefly discusses Fuzzy CNN and Fuzzy LSTM.

5.2 Fuzzification and Defuzzification

Crisp logic deals with conventional situations involving binary decisions. Examples include occurrence or non-occurrence and satisfactory or unsatisfactory. These decision-making situations are uncommon in real-world problems may be due to imprecise information. If information of this nature is used as input, the model may yield imprecise outputs (refer to Fig. 5.1); for example, in Water Distribution Networks (WDN), uncertainties associated with the cost of pipes, leakages, available heads, pipe roughness, etc., impact the design.

Fuzzy logic is viable for considering uncertainty through membership functions (MF). This process is called fuzzification (Shruti & Deka, 2020; Vasan et al., 2022). In contrast, defuzzification translates the vague form into a crisp one. Related discussion is as follows.

Two categories of MF exist for fuzzification purposes, i.e., (i) non-decreasing infers more the better and (ii) non-increasing infers less the better. The application generally guides the shape of MF. Information about selected MFs, namely, non-linear, hyperbolic, and exponential for two categories, along with related equations, are presented respectively in Figs. 5.2a, b, 5.3a, b, and 5.4a, b. Here, *Z* represents

K. Srinivasa Raju and D. Nagesh Kumar, *Artificial Intelligence and Machine Learning Techniques in Engineering and Management*,
https://doi.org/10.1007/978-981-96-2621-2_5

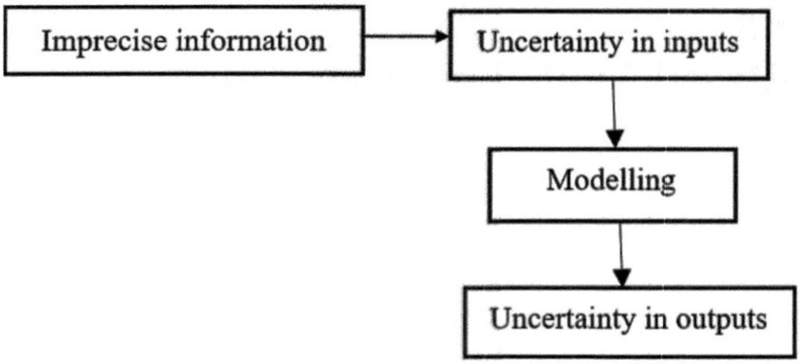

Fig. 5.1 Sources of uncertainty and their impact on output

the objective function, Z_L, and Z_U are the lowest and highest tolerance levels Z can attain (Vasan et al., 2022). Also, Fig. 5.5a, b presents the triangular and trapezoidal-shaped MFs (for the non-decreasing category), which are simple to understand and easy to interpret.

Several approaches exist for defuzzification. Some of them are weighted average and centre of gravity (Ross, 2021). Here, the centre of gravity is explained briefly due to its flexibility, which caters to most situations. The mathematical basis of the centre of gravity is (Eq. 5.1)

$$Z^* = \frac{\int_0^Z \mu_z(X) z dz}{\int_0^z \mu_z(X) dz} \quad (5.1)$$

Pictorial representation of MF for non-decreasing (or maximization) category (1)

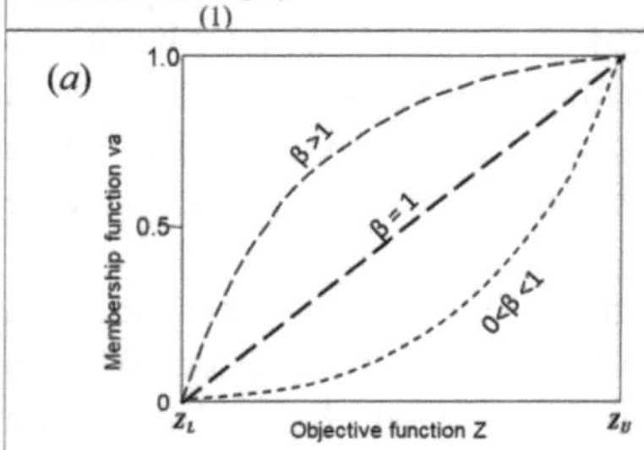

$$\mu_Z(X) = \begin{cases} 0, & Z \le Z_L \\ \left[\frac{Z - Z_L}{Z_U - Z_L}\right]^\beta, & Z_L \le Z \le Z_U \\ 1, & Z \ge Z_U \end{cases}$$

β governs the type of MF.

Pictorial representation of MF for non-increasing (or minimization) category (2)

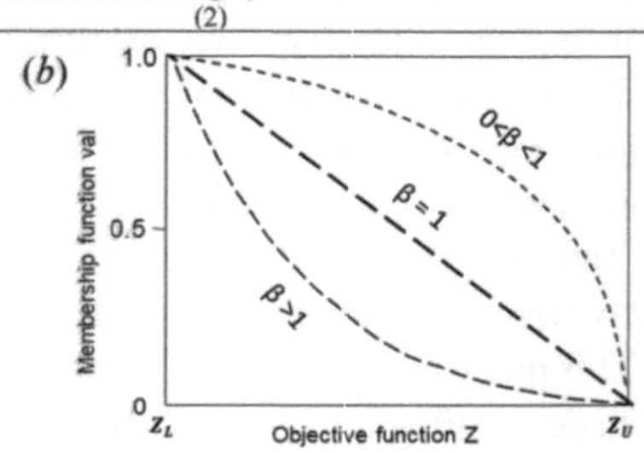

$$\mu_Z(X) = \begin{cases} 1, & Z \le Z_L \\ \left[\frac{Z_U - Z}{Z_U - Z_L}\right]^\beta, & Z_L \le Z \le Z_U \\ 0, & Z \ge Z_U \end{cases}$$

Fig. 5.2 **a, b**. Non-linear MF (Modified and adapted from Vasan et al., (2022) under CC BY-NC-ND 4.0 License)

Pictorial representation of MF for non-decreasing (or maximization) category (1)	Pictorial representation of MF for non-increasing (or minimization) category (2)

(a)

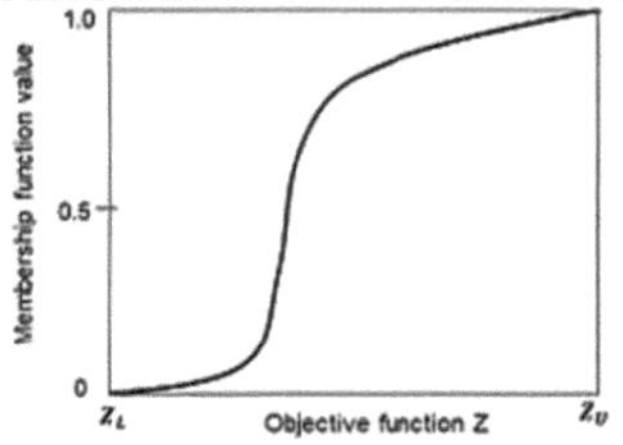

$$\mu_Z(X) = \begin{cases} 0, & Z \le Z_L \\ \frac{1}{2}tanh\left[\left(Z - \frac{Z_U + Z_L}{2}\right) \times \frac{6}{Z_U - Z_L}\right] + \frac{1}{2}, & Z_L \le Z \le Z_U \\ 1, & Z \ge Z_U \end{cases}$$

(b)

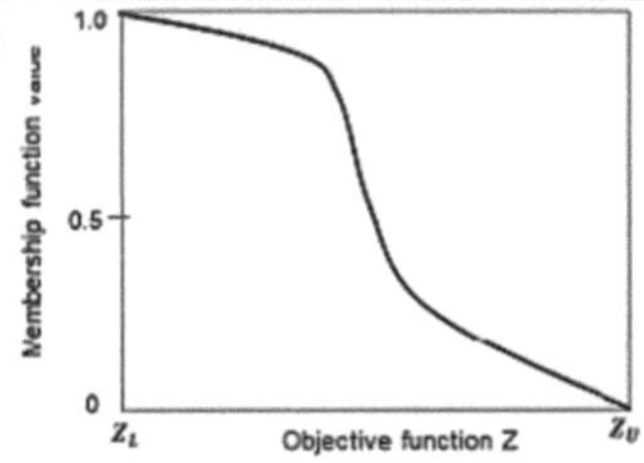

$$\mu_Z(X) = \begin{cases} 1, & Z \le Z_L \\ \frac{1}{2}tanh\left[\left(\frac{Z_U + Z_L}{2} - Z\right) \times \frac{6}{Z_U - Z_L}\right] + \frac{1}{2}, & Z_L \le Z \le Z_U \\ 0, & Z \ge Z_U \end{cases}$$

Fig. 5.3 **a** and **b**. Hyperbolic MF (Modified and adapted from Vasan et al., (2022) under CC BY-NC-ND 4.0 License)

Pictorial representation of MF for non-decreasing (or maximization) category (1)	Pictorial representation of MF for non-increasing (or minimization) category (2)

(*a*)

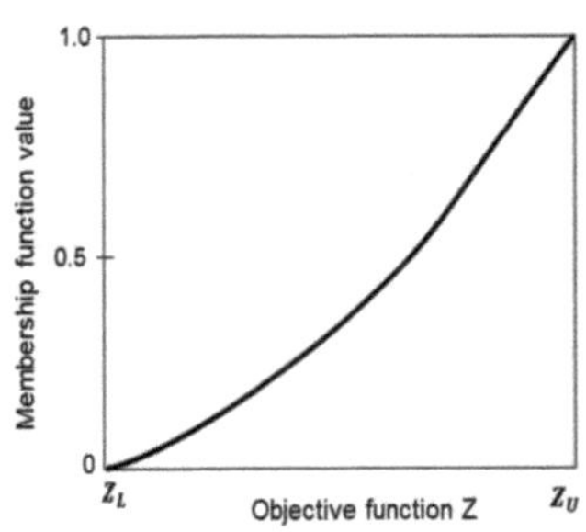

$$\mu_Z(X) = \begin{cases} 0, & Z \le Z_L \\ \left[\frac{e^{-S\left(\frac{Z_U - Z}{Z_U - Z_L}\right)} - e^{-S}}{1 - e^{-S}}\right], & Z_L \le Z \le Z_U \\ 1, & Z \ge Z_U \end{cases}$$

(*b*)

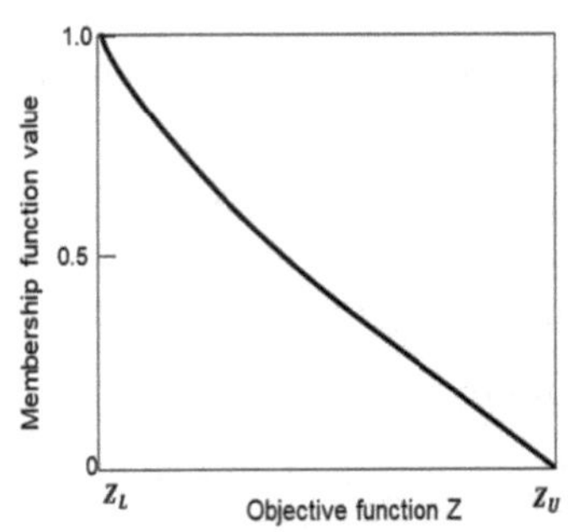

$$\mu_Z(X) = \begin{cases} 1, & Z \le Z_L \\ \left[\frac{e^{-S\left(\frac{Z - Z_L}{Z_U - Z_L}\right)} - e^{-S}}{1 - e^{-S}}\right], & Z_L \le Z \le Z_U \\ 0, & Z \ge Z_U \end{cases}$$

Fig. 5.4 **a, b** Exponential MF (Modified and adapted from Vasan et al., (2022) under CC BY-NC-ND 4.0 License)

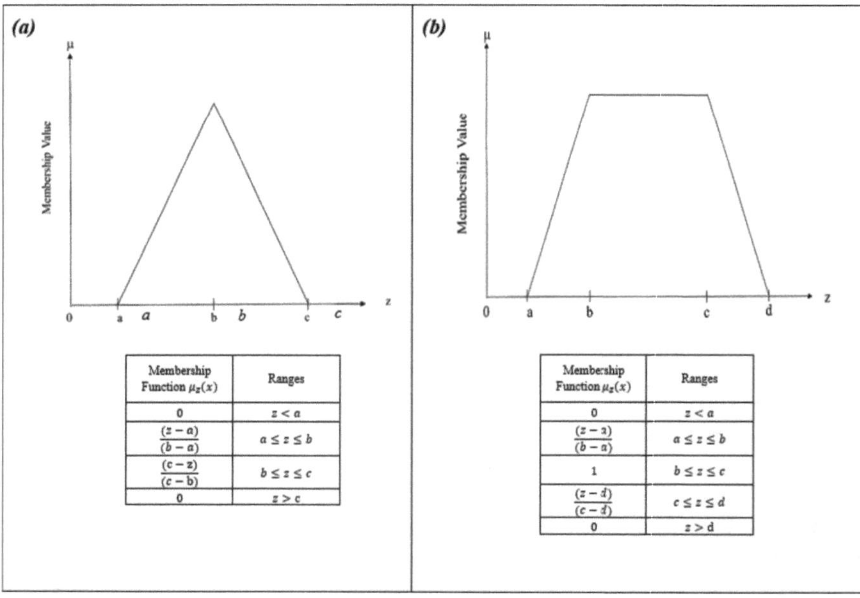

Fig. 5.5 MF and corresponding equation for **a** triangular and **b** trapezoidal

Here, $\mu_z(X)$ represents the equation of MF for the chosen line, and the defuzzified value is Z^*.

Numerical Problem 5.1. Rainfall measured using a non-recording rain gauge is 20 mm. However, after cross-verification, an error of 10% was found. Fuzzify rain gauge reading to account for this error in the triangular MF framework.

Solution:

The measured rainfall is 20 mm.

The percentage of error is 10, i.e., 0.1.

Error = 20 mm × 0.1 = 2 mm.

In this context, the lowest and highest readings that are possible are 18 mm (20–2) and 22 mm (20 + 2) (refer to Fig. 5.6).

In triangular MF, 20 mm is most likely, with a membership value of 1 (or even less in some cases, depending on the perception of the expert).

Numerical Problem 5.2. In an educational institution, students participate in several academic and non-academic activities. The institute may use student participation data to facilitate their career progression opportunities. Students with a rating score of 20 or less cannot be considered for career progression opportunities. Students above 20 and below 50 are eligible for opportunities in average category companies, whereas those above 50 and below 80 are eligible for good companies. Students with scores above 80 will be considered for the best companies. Present the data in the stepped MF framework. Assume supplementary data wherever applicable.

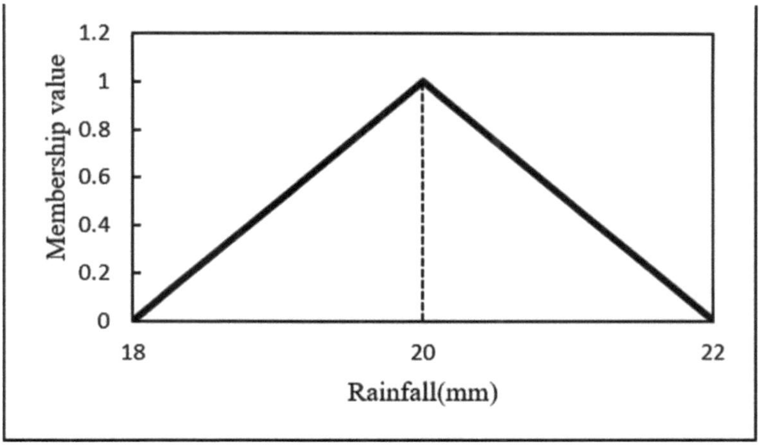

Fig. 5.6 Triangular MF for the rainfall

Solution:

It is assumed that a person with a 20-rating score or less can be given a 0.1 membership value, as they made some effort. However, they will not be eligible for career progression opportunities. A membership value of 0.4 is proposed for those above 20 and below 50 rating scores; its value is 0.7 for those above 50 and below 80 rating scores; it is suggested to be 0.9 for those above 80 (refer to Fig. 5.7).

Numerical Problem 5.3. The expert wants to understand the workability of the machines in a factory. Machines working below 10 h are considered non-productive, and machines working beyond 50 h are exceptionally productive. Draw a non-linear MF with a membership value of 0.1 for non-productive and 1 for remarkably productive. You can also provide linguistic ratings in between, such as moderately productive, and represent it on the plot.

Solution:

Machines working less than 10 h are non-productive and are assigned a value of 0.1. The logic is that it serves but does not meet the expectations of the expert. Machine working 50 h and beyond is termed exceptionally productive. Hence, a membership value of 1 is given. Here, consider 30 h for moderate productivity. You can also see this information in the plot (Fig. 5.8).

Numerical Problem 5.4. The Air Quality Index (AQI) measures pollution levels. An AQI of 100 or less is considered satisfactory, whereas an AQI of above 100 is considered not advisable. Draw a suitable MF based on your perception.

Solution:

AQI of 100 or less is considered satisfactory and can be given a membership value of 1. However, beyond 100, it is not advisable. Accordingly, a non-linear MF was

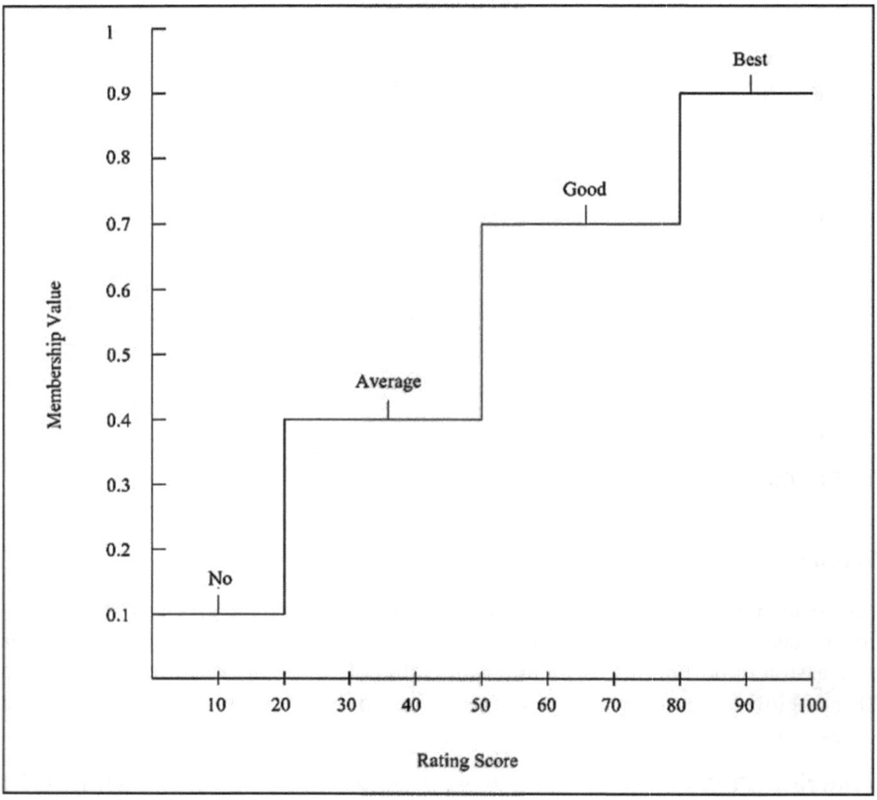

Fig. 5.7 Stepped MF for the rating score of students

proposed. With the increase in AQI, membership value is decreasing, representing not advisable [refer to Fig. 5.9].

Numerical Problem 5.5. Formulate non-decreasing exponential and hyperbolic MF in utilizing wireless sensors effectively in a highway project. The highest and lowest time sensors that can be used are 200 and 100 h. Take the value of S as 0.5.

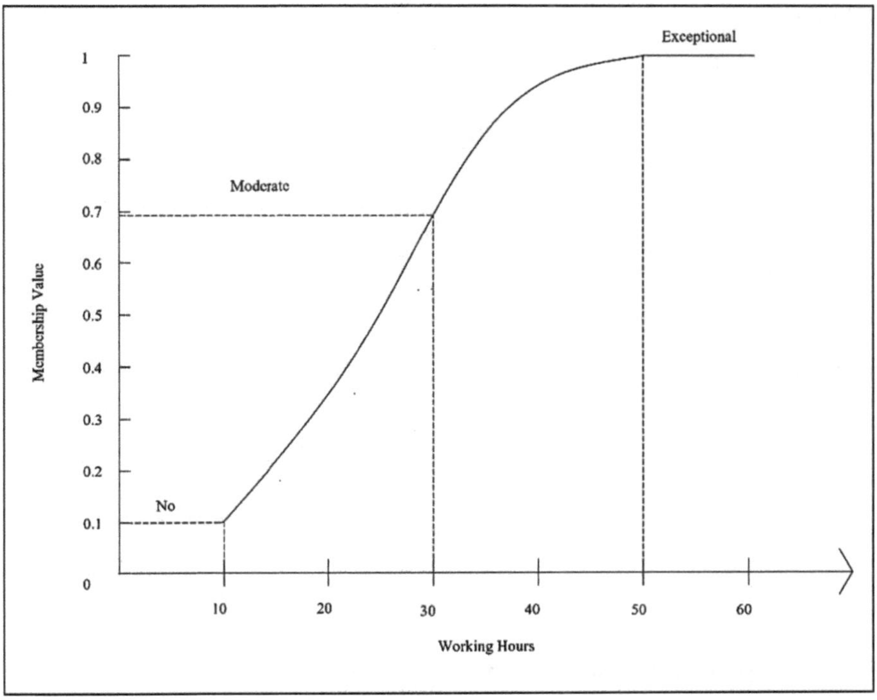

Fig. 5.8 Non-linear MF for the working hours of the machine

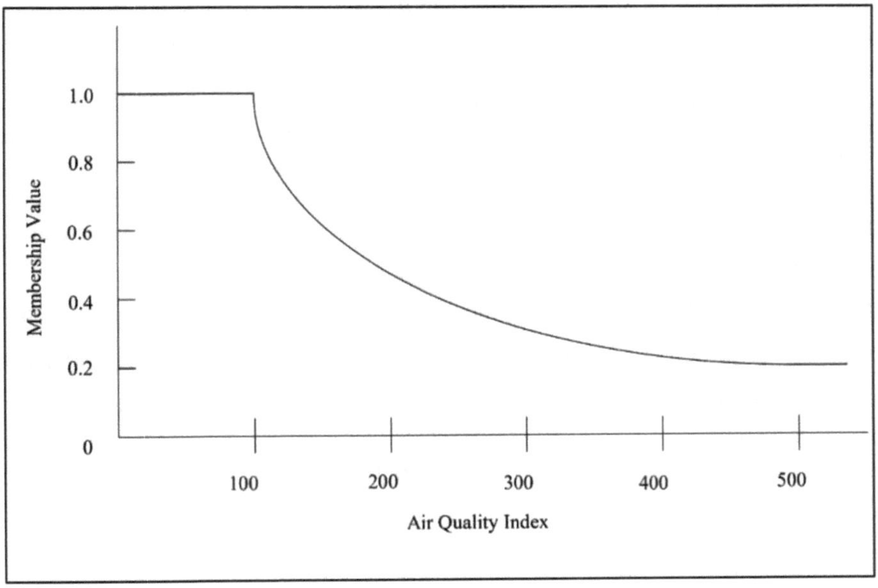

Fig. 5.9 Non-linear MF for the AQI

Solution:

Exponential MF

$$\mu_Z(X) = \left[\frac{e^{-S\left(\frac{Z_U - Z}{Z_U - Z_L}\right)} - e^{-S}}{1 - e^{-S}}\right]; Z_L \leq Z \leq Z_U$$

$$\mu_Z(X) = \left[\frac{e^{-0.5\left(\frac{200-Z}{200-100}\right)} - e^{-0.5}}{1 - e^{-0.5}}\right] = 2.5415\left[e^{\left(\frac{z-200}{200}\right)} - 0.6065\right].$$

Hyperbolic MF

$$\mu_Z(X) = \frac{1}{2}\tanh\left[\left(Z - \frac{Z_U + Z_L}{2}\right) \times \frac{6}{Z_U - Z_L}\right] + \frac{1}{2}; Z_L \leq Z \leq Z_U$$
$$= \frac{1}{2}\tanh\left[\left(Z - \frac{200 + 100}{2}\right) \times \frac{6}{200 - 100}\right] + \frac{1}{2} = \frac{1}{2}\tanh(0.06z - 9) + 0.5$$

5.3 Adaptive Neuro-Fuzzy Inference System

ANFIS simultaneously utilizes ANN and Fuzzy Inference System (FIS) (Alawad et al., 2020; Karaboga & Kaya, 2019; Larrea et al., 2021; Sada & Ikpeseni, 2021). It facilitates non-linear relationships, quick learning capability, adaptive inferences, and effectively handles noisy and inconsistent data. It collects expert knowledge and system-specific information that can be used to create fuzzy rules and MFs. Further, it can provide an excellent initial approximation, improving the overall performance.

Takagi–Sugeno, a type 3 FIS (Takagi & Sugeno, 1985), is used for demonstration where outputs are a linear combination of constant and input variables. Lastly, the weighted average of each rule outcome is determined. IF–THEN rules can be visualized as follows.

Rule 1: IF x is A_1 AND y is B_1 (antecedents), THEN $f_1 = k_1x+l_1y+m_1$ (consequent).

Rule 2: IF x is A_2 AND y is B_2 (antecedents), THEN $f_2 = k_2x+l_2y+m_2$ (consequent).

Where x and y are the inputs in the crisp set, A_i and B_i are the linguistic labels, k_i and l_i are consequent parameters, m_i is a constant, and f_i represents output MF. The typical architecture of standard ANFIS is presented in Fig. 5.10. It comprises five layers of interconnected nodes and the related discussion is explained below.

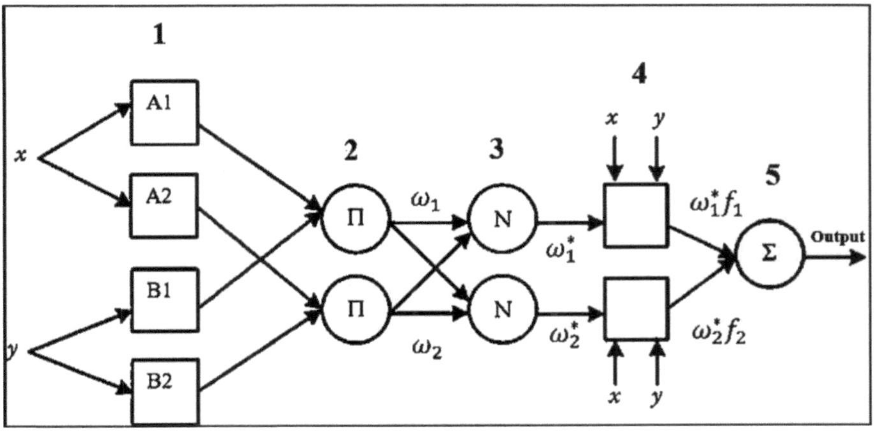

Fig. 5.10 ANFIS architecture for two features, x and y [1, 2, 3, 4, 5 are layers denoting MF, Multiplication, Normalization, Rule Functions, Summation]

Layer 1 (Fuzzification layer): Its primary role is determining the level to which a crisp input variable corresponds to a specific MF. Each node in this layer is a square shape (characterized by an adaptive node), and its output is expressed as (Eqs. 5.2–5.3):

$$Op_i^1 = \mu_{A_i}(x);\ i = 1, 2 \tag{5.2}$$

$$Op_i^1 = \mu_{B_i}(y);\ i = 3, 4 \tag{5.3}$$

Here Op_i^1 is the output from the first layer; $\mu_{A_i}(x)$, $\mu_{B_i}(y)$ are MF, respectively, for fuzzy sets A_i and B_i. The only condition is that, $\mu_{A_i}(x)$, $\mu_{B_i}(y)$ be continuous and piecewise differentiable. Changing the premise parameter would result in a different curve for the MF. The general shapes of MF are Gaussian, Trapezoidal, or Triangular. A typical Gaussian equation is presented (Eq. 5.4).

$$\mu_{A_i}(x) = e^{-\frac{1}{2}\left(\frac{x-c}{a}\right)^2} \text{ Generalized Gaussian curve} \tag{5.4}$$

Here, a and c are the standard deviation and mean, respectively.

Layer 2 (Multiplication layer): Here, nodes comprise the product of the weight of the premise parameters. It has a circle node (representing fixed node), and its output is expressed as (Eq. 5.5):

$$Op_i^2 = \omega_i = \mu_{A_i}(x) \times \mu_{B_i}(y) \tag{5.5}$$

ω_i is firing strength of ith rule. Op_i^2 represents the output of the second layer.

Layer 3 (Normalization layer): The mathematical representation of weight normalization, ω_i^N is (Eq. 5.6):

$$Op_i^3 = \omega_i^N = \frac{\omega_i}{\Sigma \omega_i} \tag{5.6}$$

where, Op_i^3 represents the output of the third layer.

Layer 4 (Defuzzification layer): It converts fuzzy information into crisp information using a defuzzification process. It consists of square nodes, and the resulting output is characterized as (Eq. 5.7).

$$Op_i^4 = \omega_i^N f_i \tag{5.7}$$

where, Op_i^4 represents the output of the fourth layer.

Layer 5 (Output layer): This process can be mathematically described as (Eq. 5.8).

$$Op_i^5 = \sum_i \omega_i^N f_i \tag{5.8}$$

where, Op_i^5 represents the output of the fifth layer.

Consequent parameters are improved (with fixed antecedent parameters) all along the forward movement, and the end output is determined. Later, the discrepancy is back propagated to the first layer, and the antecedent parameters can be improved using the chain rule (with fixed consequent parameters). Learning rate, dropout rate, batch size, epochs, and shape of MF are a few parameters that govern the ANFIS mechanism.

Chopra et al. (2021) critically summarized the advantages and disadvantages of ANFIS. High computational time, handling a considerable size of inputs (more than five), and large datasets are some challenges that affect the performance of ANFIS. Over-fitting can occur in ANFIS if the neural network component is not correctly regularized. These may require regularization methods such as weight decay and dropout. Another possibility is to explore evolutionary algorithms for training ANFIS. The workings of ANFIS are demonstrated using numerical problems.

Numerical Problem 5.6. The problem consists of inputs, Intelligence Quotient (*IQ*), and Leave in Days (*LD*). Output is Yield (Y). Data are presented in Table 5.1. Analyze the problem in the ANFIS framework.

Solution:

All the data are assumed to be normalized. Low and high linguistic levels are proposed for each input feature. Accordingly, four rules are formulated.

Table 5.1 Dataset used for ANFIS analysis

Dataset	*IQ*	*LD*	*Y*
1	104	2	4500
2	120	2	6000
3	134	1	7500
4	128	1	7000
5	130	2	6500
6	120	3	5500
7	110	2	5000
8	100	7	3500
9	110	3	4000

Rule R1: IF IQ is high AND LD is high, THEN Y is $35 \times \text{IQ} - 100 \times \text{LD} + 1000$.
Rule R2: IF IQ is high AND LD is low, THEN Y is $45 \times \text{IQ} - 450 \times \text{LD} + 2000$.
Rule R3: IF IQ is low AND LD is low, THEN Y is $35 \times \text{IQ} - 500 \times \text{LD} + 2000$.
Rule R4: IF IQ is low AND LD is high, THEN Y is $30 \times \text{IQ} - 100 \times \text{LD} + 1000$.

For the convenience of the readers, high and low are denoted as H and *L*. Accordingly, IQ high, IQ low, LD high, and LD low are now termed as IQ_H, IQ_L, LD_H, and LD_L, respectively.

Defining Gaussian MF: $e^{-\frac{1}{2}\left(\frac{x-c}{a}\right)^2}$ with basic parameters for four linguistic ratings

- IQ_L : mean $c = 105$; standard deviation $a = 10$.
- IQ_H : c and a are 130 and 10.
- LD_L : c and a are 1 and 2.
- LD_H : c and a are 6 and 2.

Demonstration of ANFIS was done in detail for rule R1 and dataset 1 for better understanding to the reader. All the results are presented in Tables 5.2, 5.3 and 5.4 for a comprehensive analysis of the numerical problem.

Layer 1: Calculating the membership values for each input of the dataset

Membership value for input (*IQ* value of 104) for linguistic rating high (with c and a are 130 and 10) is $e^{-\frac{1}{2}\left(\frac{x-c}{a}\right)^2} = e^{-\frac{1}{2}\left(\frac{104-130}{10}\right)^2} = 0.034$.

Similarly, the membership values for the remaining elements are computed based on the Gaussian MF. These values for both features are shown in Table 5.2 (columns 3–4 for *IQ* & 6–7 for *LD*).

Layer 2: Firing strength of each rule

Firing strength of rule R1 for dataset 1: Membership of IQ_H multiplied by the membership of LD_H. Here, F and M are represented for firing and MF value.

$$F(IQ_H, LD_H) = M(IQ_H) \times M(LD_H) = 0.0340 \times 0.1353 = 0.0046$$

Table 5.2 Membership values for *IQ* and *LD* for each dataset

Dataset (1)	Given Value (2)	IQ_H (3)	IQ_L (4)	Given Value (5)	LD_H (6)	LD_L (7)
1	104	0.0340	0.9950	2	0.1353	0.8825
2	120	0.6065	0.3247	2	0.1353	0.8825
3	134	0.9231	0.0149	1	0.0439	1.0000
4	128	0.9802	0.0710	1	0.0439	1.0000
5	130	1.0000	0.0439	2	0.1353	0.8825
6	120	0.6065	0.3247	3	0.3247	0.6065
7	110	0.1353	0.8825	2	0.1353	0.8825
8	100	0.0111	0.8825	7	0.8825	0.0111
9	110	0.1353	0.8825	3	0.3247	0.6065

Table 5.3 shows the firing strengths of each rule and each dataset (columns 2–5, respectively, for each rule).

Layer 3: Computation of normalized firing strength of rule R1 for dataset 1

$$
\begin{aligned}
N(IQ_H, LD_H) &= \frac{F(IQ_H, LD_H)}{\sum_{u=L}^{u=H} \sum_{v=L}^{v=H} F(IQ_u, LD_v)} \\
&= \frac{0.0046}{0.0046 + 0.0300 + 0.8781 + 0.1346} \\
&= 0.0044
\end{aligned}
$$

Here, *N* represents normalization.

Table 5.3 Firing strengths of each rule for each dataset

Dataset (1)	IQ_H,LD_H (Rule R1) (2)	IQ_H,LD_L (Rule R2) (3)	IQ_L,LD_L (Rule R3) (4)	IQ_L,LD_H (Rule R4) (5)
1	0.0046	0.0300	0.8781	0.1346
2	0.0821	0.5352	0.2865	0.0439
3	0.0405	0.9231	0.0149	0.0007
4	0.0430	0.9802	0.0710	0.0031
5	0.1353	0.8825	0.0387	0.0059
6	0.1969	0.3678	0.1969	0.1054
7	0.0183	0.1194	0.7788	0.1194
8	0.0098	0.0001	0.0098	0.7788
9	0.0439	0.0821	0.5352	0.2865

Table 5.4 Normalized firing strength of each rule for each dataset

Dataset (1)	IQ_H,LD_H (Rule R1) (2)	IQ_H,LD_L (Rule R2) (3)	IQ_L,LD_L (Rule R3) (4)	IQ_L,LD_H (Rule R4) (5)
1	0.0044	0.0286	0.8384	0.1285
2	0.0866	0.5647	0.3023	0.0463
3	0.0414	0.9427	0.0152	0.0007
4	0.0392	0.8933	0.0647	0.0028
5	0.1274	0.8307	0.0364	0.0056
6	0.2271	0.4242	0.2271	0.1216
7	0.0177	0.1153	0.7518	0.1153
8	0.0123	0.0001	0.0123	0.9753
9	0.0463	0.0866	0.5647	0.3023

Table 5.4 shows the normalized firing strengths of each rule and each dataset (columns 2–5, respectively, for each rule)

Layer 4: Multiplying the firing strength of each rule with the corresponding consequent part of that rule:

$$\begin{aligned} O_p(IQ_H, LD_H) &= N(IQ_H, LD_H) \times Consequent(IQ_H, LD_H) \\ &= 0.0044 \times (35 \times IQ_H - 100 \times LD_H + 1000) \\ &= 0.0044 \times (35 \times 104 - 100 \times 2 + 1000) \\ &= 19.54 \end{aligned}$$

Layer 5: Summing up the predicted value O_p from each rule and each dataset to get the final predicted output P and presented in Table 5.5 (columns 2–5 for prediction output for each rule). The total predicted output is given in column 6 of Table 5.5, whereas observed output and discrepancy are shown in columns 7 and 8.

$$P = \sum_{u=L}^{u=H} \sum_{v=L}^{v=H} O_p(IQ_u, LD_v) = 19.54 + 165.31 + 3890.18 + 503.72 = 4578.75$$

Numerical Problem 5.7. Engine Power (*EP*) and Miles Per Gallon (*MPG*) are influencing Fuel Consumption (*FC*) (Table 5.6). Analyze the problem in the ANFIS framework.

Table 5.5 Predicted outputs for each rule and each dataset

Dataset (1)	IQ_H, LD_H (Rule R1) (2)	IQ_H, LD_L (Rule R2) (3)	IQ_L, LD_L (Rule R3) (4)	IQ_L, LD_H (Rule R4) (5)	Predicted output considering all rules *P* (6)	Observed output *O* (7)	Discrepancy (observed–predicted) (8)
1	19.54	165.31	3890.18	503.72	4578.75	4500	−78.75
2	433.00	3670.55	1571.96	203.72	5879.23	6000	120.77
3	231.43	7145.67	94.09	3.44	7474.63	7500	25.37
4	210.90	6530.02	386.91	13.27	7141.1	7000	−141.1
5	681.59	5773.37	202.02	26.32	6683.3	6500	−183.3
6	1112.79	2566.41	1067.37	522.88	5269.45	5500	230.55
7	82.31	697.57	3646.23	472.73	4898.84	5000	101.16
8	46.74	0.34	24.60	3218.49	3290.17	3500	209.83
9	210.67	484.96	2456.45	1209.20	4361.28	4000	−361.28

Table 5.6 Dataset with two inputs and one output

Dataset	*EP*	*MPG*	*FC*
1	150	30	3100
2	120	25	400
3	180	20	6000
4	130	28	1500
5	110	22	200
6	90	18	10
7	200	22	6000
8	160	25	4000
9	120	15	750

Solution:

All the data are assumed to be normalized. Low and high linguistic levels are proposed for each input. Accordingly, four rules are formulated.

Rule R1: IF *EP* is high AND *MPG* is high, THEN *FC* is $35 \times EP - 50 \times MPG + 500$.
Rule R2: IF *EP* is high AND *MPG* is low, THEN *FC* is $40 \times EP - 150 \times MPG + 2000$.
Rule R3: IF *EP* is low AND *MPG* is low, THEN *FC* is $30 \times EP - 100 \times MPG + 1500$.
Rule R4: IF *EP* is low AND *MPG* is high, THEN *FC* is $25 \times EP - 50 \times MPG + 800$.

For the convenience of the readers, high and low are denoted as *H* and *L*. Accordingly, EP high, EP low, MPG high and MPG low are now termed as EP_H, EP_L, $\mathrm{MPG_H}$ and MPG_L, respectively.

Defining Gaussian MF: $e^{-\frac{1}{2}\left(\frac{x-c}{a}\right)^2}$ with basic parameters for four linguistic ratings.

- EP_H : mean $c = 160$; standard deviation $a = 20$.
- EP_L : c and a are 120 and 20.
- MPG_H : c and a are 25 and 5.
- MPG_L : c and a are 18 and 5.

All the results are presented in Tables 5.7, 5.8 and 5.9 for a comprehensive analysis of the numerical problem.

Layer 1: Calculating the membership values for each input of the dataset

Membership values for both inputs are shown in Table 5.7 (columns 3–4 for *EP* and 6–7 for *MPG*).

Layer 2: Firing strength of each rule

Table 5.8 shows the firing strengths of each rule for each dataset (columns 2–5, respectively, for each rule).

Layer 3: Computation of normalized firing strength.

Table 5.9 shows the normalized firing strengths of each rule for each dataset (columns 2–5, respectively, for each rule).

Layer 4: Multiplying the firing strength of each rule with the corresponding consequential part of that rule

Predicted outputs for each rule and each dataset are shown in Table 5.10 (columns 2–5).

Layer 5: Summing up the predicted value from each rule for each dataset to get the final predicted output.

The total predicted output *P* is presented in column 6 of Table 5.10, whereas observed output *O* and discrepancy are shown in columns 7 and 8. In the present case, as higher

Table 5.7 Membership values for *EP* and *MPG* for each dataset

Dataset (1)	Given value (2)	EP_H (3)	EP_L (4)	Given value (5)	MPG_H (6)	MPG_L (7)
1	150	0.8825	0.3247	30	0.6065	0.056
2	120	0.1353	1	25	1	0.3753
3	180	0.6065	0.011	20	0.6065	0.9231
4	130	0.3247	0.8825	28	0.8353	0.1353
5	110	0.044	0.8825	22	0.8353	0.7261
6	90	0.00219	0.3247	18	0.3753	1
7	200	0.1353	0.00034	22	0.8353	0.7261
8	160	1	0.1353	25	1	0.3753
9	120	0.1353	1	15	0.1353	0.8353

Table 5.8 Firing strengths of each rule for each dataset

Dataset (1)	EP_H,MPG_H (Rule R1) (2)	EP_H,MPG_L (Rule R2) (3)	EP_L,MPG_L (Rule R3) (4)	EP_L,MPG_H (Rule R4) (5)
1	0.5352	0.04942	0.0182	0.197
2	0.1353	0.05078	0.3753	1
3	0.3678	0.56	0.0102	0.0067
4	0.2712	0.044	0.1194	0.7372
5	0.0368	0.032	0.6408	0.7372
6	0.000822	0.00219	0.3247	0.1219
7	0.113	0.0982	0.000247	0.000284
8	1	0.3753	0.05078	0.1353
9	0.0183	0.113	0.8353	0.1353

Table 5.9 Normalized firing strength of each rule for each dataset

Dataset (1)	EP_H,MPG_H (Rule R1) (2)	EP_H,MPG_L (Rule R2) (3)	EP_L,MPG_L (Rule R3) (4)	EP_L,MPG_H (Rule R4) (5)
1	0.6692	0.0618	0.02276	0.2463
2	0.0867	0.0325	0.2404	0.6405
3	0.3893	0.5928	0.0108	0.0071
4	0.2314	0.0375	0.1019	0.6291
5	0.0254	0.0221	0.4429	0.5095
6	0.0018	0.0049	0.7222	0.2711
7	0.5337	0.4638	0.0012	0.0013
8	0.6405	0.2404	0.0325	0.0867
9	0.0166	0.1026	0.7581	0.1228

discrepancy values are observed, the process must be continued by changing the MF, rules, and consequent equations until a satisfactory solution is achieved.

5.4 Fuzzy Cognitive Mapping

FCM is an approach that denotes expert knowledge in a graphical network format. Nodes characterize concepts. Links symbolize association between concepts. They are instrumental in formulating the concepts and their interconnections of the edges between the nodes (Bakhtavar et al., 2021; Khanzadi et al., 2018; Papageorgiou & Salmeron, 2013). Explanation of type of relationships with two concepts, CO_i

Table 5.10 Predicted outputs for each rule for each dataset

Dataset (1)	EP_H, MPG_H (Rule R1) (2)	EP_H,MPG_L (Rule R2) (3)	EP_L,MPG_L (Rule R3) (4)	EP_L,MPG_H (Rule R4) (5)	Predicted output considering all rules P (6)	Observed output O (7)	Discrepancy (observed–predicted) (8)
1	2844.1	216.3	68.28	750.79	3060.55	3100	39.45
2	299.115	99.125	625.04	1633.11	398.24	400	1.76
3	2257.94	3675.36	52.92	30.67	5932.73	6000	67.27
4	844.61	112.5	264.94	1667.18	957.26	1500	542.74
5	82.55	68.51	1151.54	1248.45	150.83	200	49.17
6	4.95	14.21	1733.28	582.79	19.14	10	−9.14
7	3415.68	3107.46	6.36	6.22	6523.35	6000	−523.35
8	3106.425	1117.86	123.5	307.69	4223.79	4000	−223.79
9	65.57	466.83	2729.16	374.58	532.40	750	217.60

and CO_j, and the weight of interconnection between them, ω_{ij}, are presented in Table 5.11. Related details are also presented in Fig. 5.11.

Activation level Al_i for each concept CO_i is estimated using the association (Eq. 5.9).

$$Al_i^{r+1} = f\left(Al_i^r + \sum_{i=1, i \neq j}^{N} \omega_{ji} Al_j^r\right) \tag{5.9}$$

Al_i^r, Al_i^{r+1} are values of CO_i at iterations r and $r + 1$; Al_j^r is the value of CO_j at iteration r; $\omega_{ji}$is the weight of interconnection from CO_j to CO_i; f is the barrier function (Eq. 5.10)

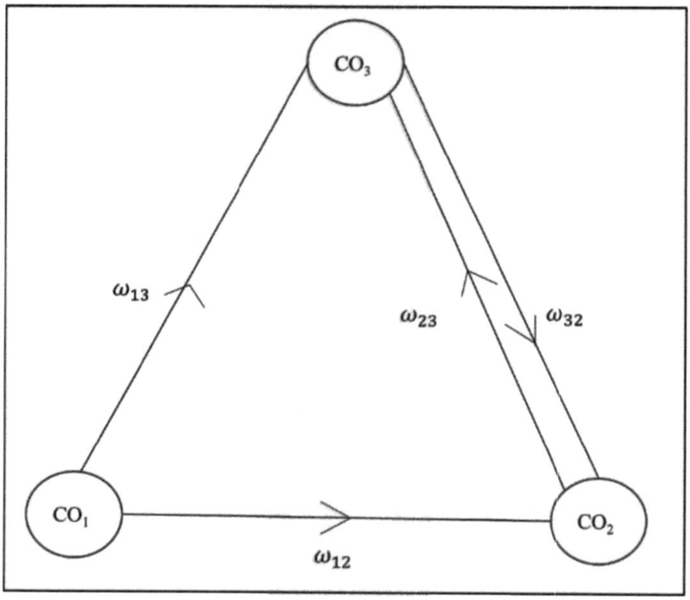

Fig. 5.11 Pictorial representation of fuzzy cognitive maps

Table 5.11 Characteristics of concepts and weights

Nature	Direction	Remark
Positive (negative)	Changes in the cause, CO_i and effect, CO_j take place in the same direction (opposite direction)	ω_{ij} has a positive sign (negative sign)
No relation	Not applicable	Edge weight zero

* Range of weights are (−1, 1)

$$f(x) = \frac{1}{1 + e^{-cx}} \tag{5.10}$$

where x and c are the input and steepness of f.

Unsupervised algorithms like Hebbian Learning (HL) improve the efficacy of FCMs by minimizing the role of expert-based knowledge (Papageorgiou & Salmeron, 2013). The procedure for training FCM is as follows (Fig. 5.12):

Here, Differential Hebbian Learning (DHL) and Non-linear Hebbian Learning (NHL) are presented, and detailed information about these is available from Papageorgiou and Salmeron (2013). In DHL, weights are updated at every iteration (Eqs. 5.11–5.12)

$$\omega_{ij}^{r+1} = \begin{cases} \omega_{ij}^{r} + L_r\left(\Delta Al_j^r \Delta Al_i^r - \omega_{ij}^{r}\right), & \Delta Al_i^r \neq 0 \\ \omega_{ij}^{r}, & \Delta Al_i^r = 0 \end{cases} \tag{5.11}$$

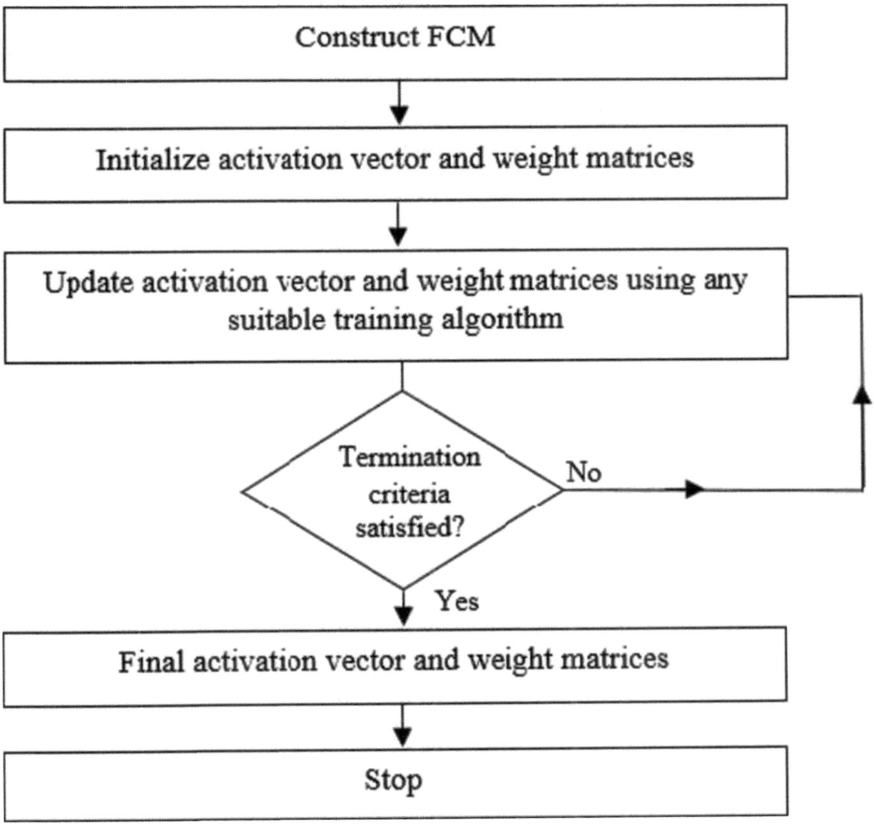

Fig. 5.12 Training process of FCM

where

$$\Delta Al_i^r = Al_i^r - Al_i^{r-1}$$

$$L_r = \left[0.1 - \frac{r}{11N}\right] \tag{5.12}$$

L_r is learning rate, N is a constant value to make sure that L_r does not take negative values during the iterative process while updating weights (Papageorgiou & Salmeron, 2013).

In NHL, weights are updated at every iteration (Eq. 5.13)

$$\omega_{ij}^{r+1} = \omega_{ij}^r + L_r Al_j^r \left(Al_i^r - \text{sgn}\left(\omega r_{ij}^r\right) Al_j^r \omega_{ij}^r\right) \tag{5.13}$$

sgn(.) is the sign function. Updation of weights continues until the termination criterion is satisfied.

Numerical Problem 5.8. There are three concepts: Terrain (1), Soil (2), and Runoff (3). Concept 1 influence 2; 2 influence 3; 3 influence 1. There is a situation that triggers concept 2. What will the effect of this situation be on all other concepts? Solve using DHL. Refer to Fig. 5.13 and Table 5.12 for the details. Take the N value as 100 and the steepness coefficient as 1.

The random weight matrix is presented in Table 5.12

The initial activation vector of the FCM would be:

$$A^0 = [0\ 1\ 0]$$

Fig. 5.13 Input data for FCM

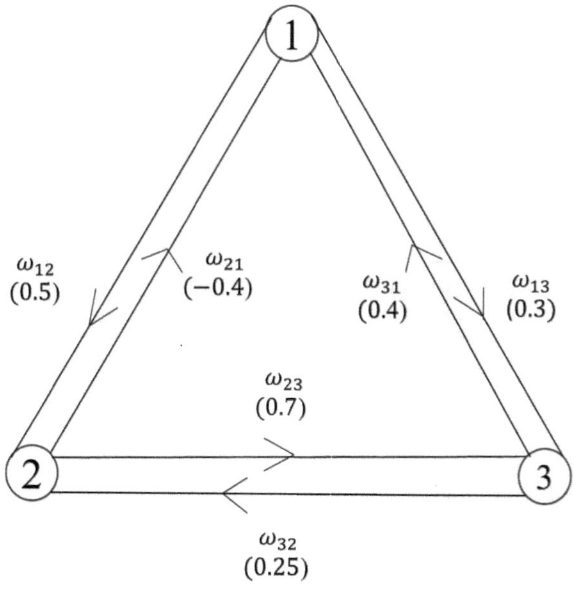

Table 5.12 Random weight matrix

Concept	1	2	3
1	0	0.5	0.3
2	−0.4	0	0.7
3	0.4	0.25	0

Solution:

Iteration $r = 1$:

The activation vector is updated using the following equation:

$Al_i^{r+1} = f\left(Al_i^r + \sum_{i=1, i \neq j}^{N} \omega_{ji} Al_j^r\right)$[refer to Eq. 5.9].

$f(x) = \frac{1}{1+e^{-x}}$; c = 1(given)

$$Al_1^1 = f(0 + 0 \times 0 + 1 \times (-0.4) + 0 \times 0.4) = f(-0.4) = \frac{1}{1 + e^{-(-0.4)}} = 0.4013$$

$$Al_2^1 = f(1 + 0 \times 0.5 + 1 \times 0 + 0 \times 0.25) = f(1) = \frac{1}{1 + e^{-1.0}} = 0.7311$$

$$Al_3^1 = f(0 + 0 \times 0.3 + 1 \times 0.7 + 0 \times 0) = f(0.7) = \frac{1}{1 + e^{-0.7}} = 0.6682$$

Hence, the activation vector will be Al^1 = [0.4013 0.7311 0.6682].

Learning rate $L_r = \left[0.1 - \frac{r}{11N}\right] = \left[0.1 - \frac{1}{1100}\right] = 0.0991$ [iteration number r = 1, N = 100].

The weight matrix is updated using Eq. (5.11), presented here again for the ready reference.

$$\omega_{ij}^{r+1} = \begin{Bmatrix} \omega_{ij}^r + L_r\left(\Delta Al_j^r \Delta Al_i^r - \omega_{ij}^r\right), & \Delta Al_i^r \neq 0 \\ \omega_{ij}^r, & \Delta Al_i^r = 0 \end{Bmatrix}$$

$$\omega_{11}^1 = 0 + 0.0991 \times ((0.4013 - 0) \times (0.4013 - 0) - 0)$$
$= 0.016 \sim 0$ (rounded to zero as the comparison is against the same concept and applies to all diagonal elements)

$$\omega_{12}^1 = 0.5 + 0.0991 \times ((0.7311 - 1) \times (0.4013 - 0) - 0.5) = 0.4398$$

$$\omega_{13}^1 = 0.3 + 0.0991 \times ((0.6682 - 0) \times (0.4013 - 0) - 0.3) = 0.2968$$

$$\omega_{21}^{1} = -0.4 + 0.0991 \times ((0.4013 - 0) \times (0.7311 - 1) - (-0.4)) = -0.3711$$

$$\omega_{22}^{1} = 0 + 0.0991 \times ((0.7311 - 1) \times (0.7311 - 1) - 0) = 0.0072 \sim 0$$

$$\omega_{23}^{1} = 0.7 + 0.0991 \times ((0.6682 - 0) \times (0.7311 - 1) - 0.7) = 0.6128$$

$$\omega_{31}^{1} = 0.4 + 0.0991 \times ((0.4013 - 0) \times (0.6682 - 0) - 0.4) = 0.3869$$

$$\omega_{32}^{1} = 0.25 + 0.0991 \times ((0.7311 - 1) \times (0.6682 - 0) - 0.25) = 0.2074$$

$$\omega_{33}^{1} = 0 + 0.0991 \times ((0.6682 - 0) \times (0.6682 - 0) - 0) = 0.0442 \sim 0$$

The updated weight matrix after iteration 1 is (refer to Table 5.13):
Iteration $r = 2$:

$$\begin{aligned} Al_1^2 &= f(0.4013 + (0.4013 \times 0) + (0.7311 \times (-0.3711)) + (0.6682 \times 0.3869)) \\ &= f(0.3885) = \frac{1}{1 + e^{-0.3885}} = 0.5959 \end{aligned}$$

$$\begin{aligned} Al_2^2 &= f(0.7311 + (0.4013 \times 0.4398) + (0.7311 \times 0) + (0.6682 \times 0.2074)) \\ &= f(1.0462) = \frac{1}{1 + e^{-1.0462}} = 0.7400 \end{aligned}$$

$$\begin{aligned} Al_3^2 &= f(0.6682 + (0.4013 \times 0.2968) + (0.7311 \times 0.6128) + (0.6682 \times 0)) \\ &= f(1.2353) = \frac{1}{1 + e^{-1.2353}} = 0.7747 \end{aligned}$$

Hence, the activation vector will be $Al^2 = [0.5959\ 0.7400\ 0.7747]$.
The learning rate for iteration 2 is $\left[0.1 - \frac{r}{11N}\right] = \left[0.1 - \frac{2}{1100}\right] = 0.0982$.
The weight matrix is updated which is as follows:

$$\omega_{11}^{1} = 0 + 0.0982 \times ((0.5959 - 0.4013) \times (0.5959 - 0.4013) - 0) = 0.00372 \sim 0$$

Table 5.13 Weight matrix after iteration 1

Impact	1	2	3
1	0	0.4398	0.2968
2	−0.3711	0	0.6128
3	0.3869	0.2074	0

$$\omega_{12}^{1} = 0.4398 + 0.0982 \times ((0.7400 - 0.7311) \times (0.5959 - 0.4013) - 0.4398) = 0.3968$$

$$\omega_{13}^{1} = 0.2968 + 0.0982 \times ((0.7747 - 0.6682) \times (0.5959 - 0.4013) - 0.2968) = 0.2697$$

$$\omega_{21}^{1} = -0.3711 + 0.0982 \times ((0.5959 - 0.4013) \times (0.7400 - 0.7311) - (-0.3711)) = -0.3345$$

$$\omega_{22}^{1} = 0 + 0.0982 \times ((0.7400 - 0.7311) \times (0.7400 - 0.7311) - 0) = 0.0000077784 \sim 0$$

$$\omega_{23}^{1} = 0.6128 + 0.0982 \times ((0.7747 - 0.6682) \times (0.7400 - 0.7311) - 0.6128) = 0.5527$$

$$\omega_{31}^{1} = 0.3869 + 0.0982 \times ((0.5959 - 0.4013) \times (0.7747 - 0.6682) - 0.3869) = 0.3509$$

$$\omega_{32}^{1} = 0.2074 + 0.0982 \times ((0.7400 - 0.7311) \times (0.7747 - 0.6682) - 0.2074) = 0.1871$$

$$\omega_{33}^{1} = 0 + 0.0982 \times ((0.7747 - 0.6682) \times (0.7747 - 0.6682) - 0) = 0.001113 \sim 0$$

The updated weight matrix after iteration 2 is presented in Table 5.14.

Numerical Problem 5.9. Solve numerical problem 5.8 using NHL (refer to Table 5.12). The learning rate is 0.001. Show computations for one iteration.

Solution:

Iteration $r = 1$:

The activation vector is updated

$$A_{l1}^{1} = f(0 + 0 \times 0 + 1 \times (-0.4) + 0 \times 0.4) = f(-0.4) = \frac{1}{1 + e^{-(-0.4)}} = 0.4013$$

$$Al_2^{1} = f(1 + 0 \times 0.5 + 1 \times 0 + 0 \times 0.25) = f(1) = \frac{1}{1 + e^{-1.0}} = 0.7311$$

$$Al_3^{1} = f(0 + 0 \times 0.3 + 1 \times 0.7 + 0 \times 0) = f(0.7) = \frac{1}{1 + e^{-0.7}} = 0.6682$$

Hence, the activation vector will be $Al^{1} = [0.4013\ 0.7311\ 0.6682]$.

The weight matrix is updated using Eq. 5.13

$$\omega_{ij}^{r+1} = \omega_{ij}^{r} + L_r Al_j^{r}\left(Al_i^{r} - \mathrm{sgn}\left(\omega r_{ij}^{r}\right) Al_j^{r} \omega_{ij}^{r}\right)$$

Table 5.14 Weight matrix after iteration 2

Impact	1	2	3
1	0	0.3968	0.2697
2	−0.3345	0	0.5527
3	0.3509	0.1871	0

$$\omega_{11}^{1} = 0 + 0.001 \times 0.4013 \times (0.4013 - 0.4013(0)) = 0.00016 \sim 0$$

$$\omega_{12}^{1} = 0.5 + 0.001 \times 0.7311 \times (0.4013 - 0.7311 \times 0.5) = 0.5$$

$$\omega_{13}^{1} = 0.3 + 0.001 \times 0.6682 \times (0.4013 - 0.6682 \times 0.3) = 0.3001$$

$$\omega_{21}^{1} = -0.4 + 0.001 \times 0.4013 \times (0.7311 - (-)0.4013 \times (-0.4) = -0.3998$$

$$\omega_{22}^{1} = 0 + 0.001 \times 0.7311 \times (0.7311 - 0.7311 \times 0) = 0.00053 \sim 0$$

$$\omega_{23}^{1} = 0.7 + 0.001 \times 0.6682 \times (0.7311 - 0.6682 \times 0.7) = 0.7002$$

$$\omega_{31}^{1} = 0.4 + 0.001 \times 0.4013 \times (0.6682 - 0.4013 \times 0.4) = 0.4002$$

$$\omega_{32}^{1} = 0.25 + 0.001 \times 0.7311 \times (0.6682 - 0.7311 \times 0.25) = 0.2504$$

$$\omega_{33}^{1} = 0 + 0.001 \times 0.6682 \times (0.6682 - 0.6682 \times 0) = 0.00045 \sim 0$$

The weight matrix after iteration 1 is (refer to Table 5.15):

5.5 Fuzzy Logic-Based Optimization

Optimization techniques play a significant role in engineering and management, where there are recurrent phenomena of resource limitation and massive requirements. The perennial question among policymakers is how best to utilize the available resources with the existing constraints to maximize achievements (Loucks & Beek, 2017). Components that govern the workflow of the optimization process are described as follows:

- Decision variables (DV) are the variables set that controls the problem.
- The objective function (O) represents the goal of the problem.

Table 5.15 Weight matrix after iteration 1

Impact	1	2	3
1	0	0.5	0.3001
2	−0.3998	0	0.7002
3	0.4002	0.2504	0

- Constraints are the challenges that do not allow the objective to achieve its full potential.
- Bounds: Allow the unknown decision variables to take on specific values within a range.
- Representative solution techniques for obtaining the optimal solutions are Linear Programming (LP), Non-linear Programming (Non-LP), and Quadratic Programming (QP).
- Optimum output: DV obtained after optimization and related objective function

The process also involves extensive data collection, which may have to be refined before using as input to the optimization model. A mathematical description is presented in Table 5.16 with three DVs, x_1, x_2, x_3, with an intent to demonstrate chosen optimization techniques (Rao, 2013).

As discussed in Table 5.16, *SS* depends on *O* and *CS* in any optimization framework. In a fuzzy context, these are expressed as (Gaur et al., 2015; Morankar et al., 2016; Vasan et al., 2022) (Eq. 5.14):

$$\mu_{SS} = (\mu_O \cap \mu_{CS}) \tag{5.14}$$

Table 5.16 Example for demonstrating chosen optimization techniques

Characteristic	LP	Non-LP	QP
Objective function	Linear functions of DV	Non-linear and linear functions of DV	Quadratic and linear functions of DV
Constraints	Linear functions of DV	Non-linear and linear functions of DV	Linear functions of DV
Mathematical representation of O	Max/Min $1600\,x_1 + 1700\,x_2 + 1800\,x_3$	Max/Min $1600\,x_1^2 + 1700\,x_2^{1.8} + 1800\,x_3^{4.2}$	Max/Min $1600\,x_1^2 + 1700\,x_2^2 + 1800\,x_3^2 + 1800\,x_1 + 1900\,x_2 + 2000\,x_3$
Mathematical representation of constraints (CS)	$0.06\,x_1 + 0.16\,x_2 + 0.18\,x_3 \le 6$ $x_1 + x_2 + x_3 \le 400$	$0.06\,x_1^{4.2} + 0.16\,x_2^{5.4}x_2^2 + 0.12\,x_3^{2.6} \le 6$ $x_1^{2.8} + x_2 + x_3^{4.4} \le 400$	$0.06\,x_1 + 0.16\,x_2 + 0.12\,x_3 \le 6$ $x_1 + x_2 + x_3 \le 400$
Bounds [Assuming linear variation of bounds]	$20 \le x_1 \ge 50$ $30 \le x_2 \ge 60$ $40 \le x_3 \ge 80$	$20 \le x_1 \ge 50$ $30 \le x_2 \ge 60$ $40 \le x_3 \ge 80$	$20 \le x_1 \ge 50$ $30 \le x_2 \ge 60$ $40 \le x_3 \ge 80$
Decision space (*SS*), based on O and *CS*	Global optimum solution	It is likely the local (or global) optimal solution	
Remarks	No uncertainty in parameters, objective function, or constraints is considered		

$\mu_{SS}, \mu_O, \mu_{CS}$ represent MF corresponding to SS, O, and CS. With several objective functions (1, 2, n) and constraints (1, 2, m), Eq. 5.14 can be transformed into Eqs. (5.15–5.16)

$$\mu_{SS}(X) = \left[\mu_{O1}(X) \cap \mu_{O2}(X) \cap \ldots \cap \mu_{On}(X) \cap \mu_{CS1}(X) \cap \mu_{CS2}(X) \cap \ldots \cap \mu_{CSm}(X)\right] \tag{5.15}$$

Here, AND denotes intersection (∩), i.e., minimum (Morankar et al., 2013; Zimmermann, 1991)

$$\mu_{SS}(X) = Min[\mu_{O1}(X), \mu_{O2}(X), \ldots, \mu_{On}(X), \mu_{CS1}(X), \mu_{CS2}(X), \ldots, \mu_{CSm}(X)] \tag{5.16}$$

The optimum solution is (Eq. 5.17)

$$\mu^*_{SS}(X) = Max[(\mu_{SS}(X)] \tag{5.17}$$

Degree of satisfaction λ, an auxiliary continuous variable, is introduced as an equivalent optimization problem in the single objective framework (Eq. 5.18). The intention is to identify a unique solution x*, which facilitates the optimum output (in this case λ) (Lence et al., 2017; Sasikumar & Mujumdar, 1998) (Eqs. 5.18–5.21)

$$Max\ \lambda \tag{5.18}$$

subject to

$$\mu_{Oj}(X) \geq \lambda \quad j = 1, 2, \ldots, n \tag{5.19}$$

$$\mu_{CSi}(X) \geq \lambda \quad i = 1, 2, \ldots, m \tag{5.20}$$

$$0 \leq \lambda \leq 1 \tag{5.21}$$

In addition, all other case study-related constraints and bounds must be considered.

However, the intensity of high computational requirements in the case of traditional non-linear optimization techniques motivated the search for new approaches. In this regard, evolutionary optimization algorithms have gained prominence for solving complex problems (Reddy & Kumar, 2020) and are briefly discussed in Chap. 6.

Numerical Problem 5.10. Four categories of machines are proposed to be installed in a workshop where space is available for 15 machines. The lubricating oil required for maintenance of each type of machine is 0.1 L (Here L is Litre), 0.2 L, 0.4 L, and 0.1 L, whereas available is 4.5 L. The working capacity of one unit of each category machine for a day is 3, 5, 5, and 3 h. The minimum number of machines expected to be installed is 1, 2, 4, and 2; the maximum is 2, 4, 5, and 7. Mention the decision variables. Formulate the problem for maximization of the working capacity of machines and solve it in an LP framework.

Solution:

Let m_1, m_2, m_3, m_4 are the number of machines proposed under four different categories in a workshop and these are the decision variables.

Objective function is the maximization of the working capacity of machines termed *WC*

$$Max\ WC = 3m_1 + 5m_2 + 5m_3 + 3m_4$$

Subjected to:

Constraints

$$m_1 + m_2 + m_3 + m_4 \leq 15$$

$$0.1m_1 + 0.2m_2 + 0.4m_3 + 0.1m_4 \leq 4.5$$

Bounds

$$\begin{aligned} 1 \leq m_1 \leq 2 \\ 2 \leq m_2 \leq 4 \\ 4 \leq m_3 \leq 5 \\ 2 \leq m_4 \leq 7 \end{aligned}$$

Solution:

The optimal working capacity of machines is 63 h, $\begin{aligned} m_1 &= 1; \\ m_2 &= 4; \\ m_3 &= m_4 = 5 \end{aligned}$

Numerical Problem 5.11. Solve the following two objectives (maximizing the water supply and minimizing aquifer loss) problem in a fuzzy non-linear optimization framework.

$Max\, z_1 = 320x_1 + 440x_2 + 620x_3$ [maximizing the water supply]

$Min\, z_2 = 6x_1 + 2x_2 + 3_3$ [minimizing the aquifer loss]

subject to

$$0.2x_1 + 0.28x_2 + 0.6x_3 \leq 42$$

$$0.26x_1 + 0.44x_2 \leq 24$$

$$20 \leq x_1 \leq 44;$$

$$30 \leq x_2 \leq 48$$

$$25 \leq x_3 \leq 45$$

Use non-linear MF with $\beta = 3$. Consider uncertainty in objective functions only.

Solution:

Maximization and minimization of each objective function provide higher and lower limits, respectively (Columns 2–5 for individual higher and lower values for each objective, Table 5.17).

The non-linear MF-based optimization model is as follows:

Max λ

subject to

Table 5.17 Results of fuzzy optimization

Characteristics (1)	Higher Z_1 (2)	Lower Z_1 (3)	Higher Z_2 (4)	Lower Z_2 (5)	Solution with non-linear MF (both objectives $\beta = 3$) (6)
x_1	41.53846	20	41.53846	20	20
x_2	30	30	30	30	42.72727
x_3	42.15385	25	42.15385	25	35.38240
Objective function value	52,627.7	35,100	435.69	255	(Maximum 47,137.09, minimum 311.6)
					Optimum degree of satisfaction λ = 0.32388

$$\mu_Z(X) = \left[\frac{Z - Z_L}{Z_U - Z_L}\right]^\beta = \left[\frac{320x_1 + 440x_2 + 620x_3 - 35100}{52627.7 - 35100}\right]^3 \geq \lambda$$

$$\mu_Z(X) = \left[\frac{Z_U - Z}{Z_U - Z_L}\right]^\beta = \left[\frac{435.69 - (6x_1 + 2x_2 + 3x_3)}{435.69 - 255}\right]^3 \geq \lambda$$

$$0.2x_1 + 0.28x_2 + 0.6x_3 \leq 42$$

$$0.26x_1 + 0.44x_2 \leq 24$$

$$x_1 \geq 20; x_1 \leq 44;$$

$$x_2 \geq 30; x_2 \leq 48$$

$$x_3 \geq 25; x_3 \leq 45$$

Related results of Non-Linear Optimization are (Column 6 of Table 5.17):

- Maximum λ is 0.32388, and the corresponding (x_1, x_2, x_3) values are (20, 42.72727, and 35.3824).
- The solution obtained by the fuzzy optimization problem is between the lowest and highest values obtained by individual objectives (columns 2–5), representing a compromise solution with the tradeoff of water supply and aquifer loss.

5.6 Fuzzy CNN, Fuzzy LSTM, and Fuzzy CNN-LSTM

Fuzzy CNN employs a fuzzy inference layer in place of a fully connected to integrate the features more effectively (Lin & Jhang, 2022). The mathematical philosophy of Fuzzy CNN till the flattened layer remains the same compared to CNN (Fig. 5.14). The fuzzy inference layer is explained mathematically in terms of the symmetric Gaussian membership function using linguistic terms (High, Medium, Low; H, M, L). These fuzzified features are further used to create rules R. The initial membership function $\mu_{F_i^q}$, for qth output inference having feature maps p_f is calculated for each input feature and directly used in the estimation of the output $\emptyset_q^{p_f}$ (Langeroudi et al., 2022). Hereafter, combined weighted inference (z) is obtained through the multiplication of weight (ω) and inference ($\emptyset$) matrices. Further, the z is normalized to get output. Note that the output changes when different activation functions are considered. Back-propagation of the error is accomplished by improving the gradient values (Hsu et al., 2020).

Fuzzy LSTM is an architecture of LSTM in a fuzzy framework (Li et al., 2020). Fuzziness is incorporated through the fuzzy inference layer immediately after the

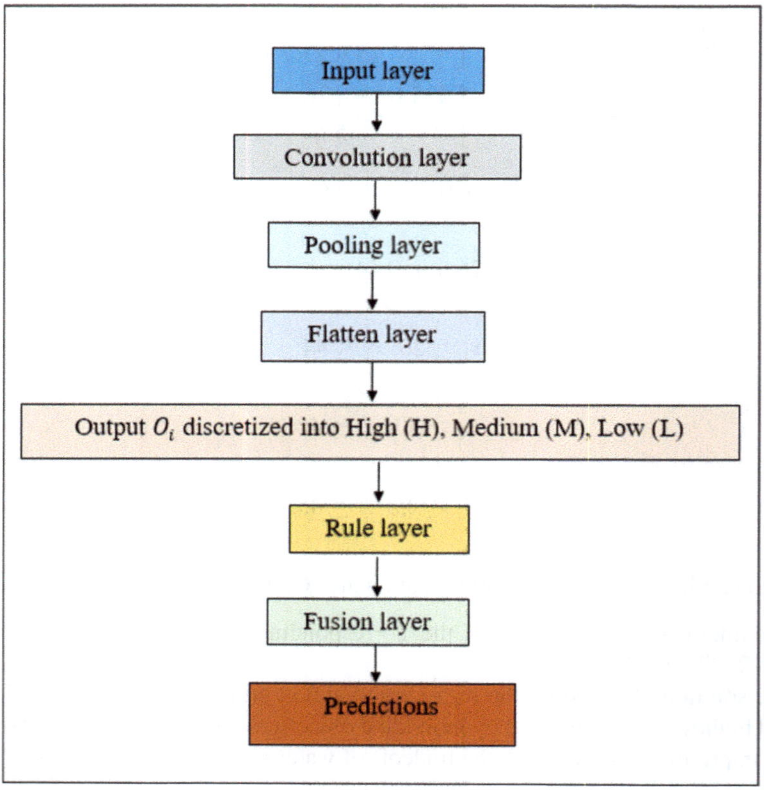

Fig. 5.14 The architecture of fuzzy CNN

output of LSTM (Fig. 5.15). This type of incorporation is done to avoid disrupting information flow between the forget, input, and output gates. These gates maintain the balance between retaining helpful information and discarding irrelevant information. Consequently, adding fuzziness between these gates potentially reduces the ability of the algorithm to capture long-term dependencies. Therefore, applying fuzziness after the output stage helps maintain model interpretability, handles uncertainties, and ensures the LSTM's core operations remain intact (Langeroudi et al., 2022).

The fuzzy inference layer mainly comprises fuzzy sets, μ, tensor layer (T_l), rule layer (R), and fusion layer (F). μ, R, and Ø are the same as Fuzzy-CNN. The fuzzified features obtained from the μ forms T_l. After that, the rules are created on the features and the μ. The final layer is the F which employs three operations:

i. μ and R are connected utilizing the concatenation operation and are characterized by $F(\mu, R)$.
ii. It undergoes a linear transformation.
iii. Subsequently, a selected activation function is applied to achieve the desired outcome.

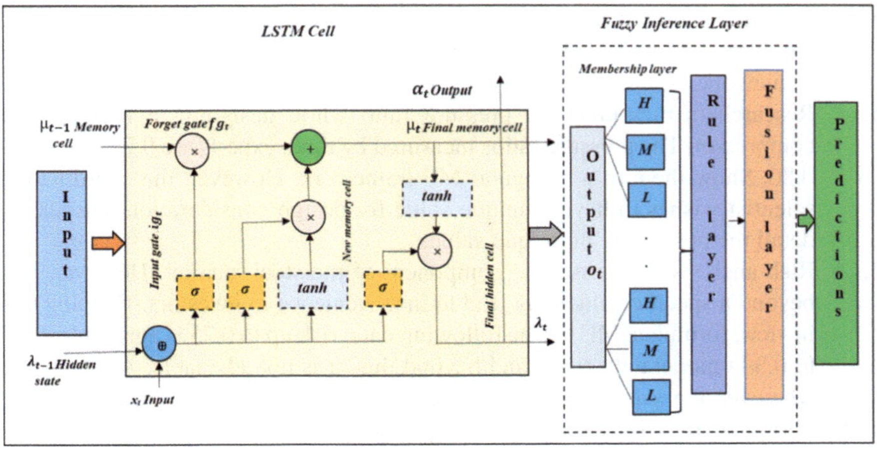

Fig. 5.15 The architecture of fuzzy LSTM (Modified and adapted from Vogeti et al., 2024 under CC BY-NC-ND 4.0 License)

Extensive details of these techniques are available from Vogeti et al. (2024).

Fuzzy CNN-LSTM is an extension of CNN-LSTM in the fuzzy framework (Bao et al., 2022).

Representative Software

ANFIS in MatLab perspective [https://www.mathworks.com/help/fuzzy/neuro-adaptive-learning-and-anfis.html, accessed on 07.04.2023].

Fuzzy Cognitive Mapping: Mental Modeler https://www.mentalmodeler.com/, accessed on 07.04.2023].

LINGO handles Linear, Non-linear based optimization problems https://www.lindo.com/index.php/products/lingo-and-optimization-modeling

Revision Questions and Exercise Problems

5.1 Differentiate crisp and fuzzy logic.
5.2 What are the causes of uncertainty?
5.3 What are the associated uncertainties in WDN?
5.4 What is an MF? What is its purpose?
5.5 What are fuzzification and defuzzification?
5.6 What are the possible shapes of MF?
5.7 What is the meaning of non-increasing and non-decreasing MF?
5.8 What is the significance of β in the context of non-linear fuzzy optimization?
5.9 What is the significance of S in the case of the exponential MF?
5.10 What is the mathematical philosophy of triangular and trapezoidal MF?
5.11 Discuss the Centre of gravity-based defuzzification method.

5.12 Effective sunshine hours in the semi-arid zone are recorded as 30% of the day. Later, it was found to have an error of 15%. Fuzzify in a triangular MF framework.
5.13 Researchers measured the pressure ratio while designing a fuel cell air compressor. The pressure ratios measured by three experts are 0.72, 0.76, and 0.80. Show them in a triangular MF framework. However, the coordinating scientist wishes to have a unique value for design consideration. Discuss in detail the possibility of unique value.
5.14 Risk analysis is an essential component of the stock market. However, risk beyond a specific value may lead to inconvenience to investors. Keeping this in view, formulate MF for the following data: risk up to 0.25 is agreeable; 0.25 to 0.3 is moderately risky, and beyond this, it is not advisable. Assume data wherever applicable. Show the same in the appropriate MF framework.
5.15 Formulate non-decreasing exponential and hyperbolic MF for the data related to energy management. The highest value of an objective function Z is 350, whereas the lowest is 180. Take the value of exponential parameter S as 0.8.
5.16 In a transporation economics problem, monetary benefits play a significant role and are expressed as $18x_1 + 8x_2$. Here, x_1 and x_2 are governing decision variables. The highest and lowest values of the objective function are 200 and 140 units. The value of β is 3.4. Express the problem in the non-linear MF format.
5.17 What is ANFIS?
5.18 How many layers exist in ANFIS? Explain their functionality in brief.
5.19 What is the mathematical expression of the Gaussian MF?
5.20 Analyze the given problem from an ANFIS perspective. Use the data from Table 5.1.

IF IQ is *high* AND LD is *high*, THEN Y is $50 \times IQ - 225 \times LD + 1800$
IF IQ is *high* AND LD is *low*, THEN Y is $38 \times IQ - 590 \times LD + 1800$
IF IQ is *low* AND LD is *low*, THEN Y is $55 \times IQ - 600 \times LD + 1560$
IF IQ is *low* AND LD is *high*, THEN Y is $30 \times IQ - 140 \times LD + 600$

5.21 What are the concepts and weights in the case of FCM?
5.22 What is the range of weights in the case of FCM?
5.23 What is an activation vector? What is its purpose?
5.24 What is the purpose of HL algorithms?
5.25 Is there a reduction in the dependency of FCM on expert's knowledge using HL algorithms?
5.26 Do HL algorithms fall under supervised or unsupervised approaches?
5.27 What are the different types of HL algorithms?
5.28 What are the parameters that affect learning rates in the case of DHL?
5.29 What is the significance of the learning rate in HL algorithms?
5.30 How does weight updation differ in DHL and NHL?
5.31 Which has the least computational complexity, DHL or NHL? Justify?
5.32 Three concepts, mental health MH, classroom factors CF, and socioeconomic status SS, impact students' behaviour. MH impacts CF; CF impacts SS; SS

Table 5.18 Random weight matrix

Impact	MH	CF	SS
MH	0	0.28	0
CF	0.5	0	0.6
SS	0.80	0.3	0

impacts MH. There is a situation that triggers the concept of MH. How will this situation affect all other concepts? Solve using DHL and NHL. Table 5.18 shows a random weight matrix.

The initial activation vector is [1, 0, 0]. Assume suitable values while solving the problem.

5.33 What is the physical interpretation of decision variables in the optimization framework?
5.34 What is the philosophy of objective functions and constraints?
5.35 What is the difference between constraints and bounds?
5.36 What is the difference between output and optimum output?
5.37 What are the possible objective functions in the case of WDN?
5.38 What is the workflow while solving optimization problems?
5.39 What is the mathematical difference between LP and QP?
5.40 What is the mathematical difference between Non-LP and QP?
5.41 What is the necessity of fuzzy optimization? How is it different from crisp optimization?
5.42 On what parameters does decision space depend?
5.43 What is an auxiliary variable in the context of fuzzy optimization?
5.44 What is the degree of satisfaction? What is its purpose in the case of fuzzy optimization?
5.45 Can the degree of satisfaction be considered as an objective function in fuzzy optimization? Discuss in detail.
5.46 Do you prefer a higher degree of satisfaction or a lower one?
5.47 What is the range of degree of satisfaction? What is the physical significance if it is 1?
5.48 Formulate an optimization problem in a fuzzy optimization framework using non-linear MF with $\beta = 6$, exponential MF with $S = 0.8$, and hyperbolic MF from your domain of interest. Consider uncertainty in objective functions only. You are expected to take three decision variables and two objective functions with maximization in nature. You can think of keeping three constraints.
5.49 What is the difference between fuzzy CNN and CNN? Discuss in detail.
5.50 What is the difference between fuzzy LSTM and LSTM? Discuss in detail.

Advanced Review Questions

5.51 Can you identify four situations in which MF can be employed?
5.52 Mention one situation in your domain of interest to discuss unquantifiable, non-obtainable, incomplete information. Relate with examples.

5.53 Relate hyperbolic and exponential MF with one example in your domain of interest.
5.54 Can you propose two new MFs with corresponding mathematical equations?
5.55 Do you prefer a triangular or trapezoidal MF? Why?
5.56 Several defuzzification methods exist. Analyze and compare the same.
5.57 What is FIS? Does it work on a rule-based platform? If yes, expand your answer. If not, explain the logic of the same.
5.58 How Mamdani and Sugeno FIS differ? Explain mathematically.
5.59 What is the purpose of antecedent and consequent parameters in FIS?
5.60 Discuss three case studies related to Mamdani FIS.
5.61 Discuss how Mamdani's approach can be facilitated in any programming platform.
5.62 Discuss three case studies related to ANFIS.
5.63 Discuss in detail how ANFIS can be facilitated in any programming platform.
5.64 How does FCM help the decision-making process in your research area? Discuss with related case studies.
5.65 Mention four examples of optimization in your research area. Also, mention decision variables, objective functions, and constraints.
5.66 Are fuzzy optimization approaches capable of handling the uncertainty in the data?
5.67 Discuss two case studies where fuzzy optimization was employed. Emphasize the decision variables, objective function, constraints, and bounds.
5.68 Can you modify the existing architectures of Fuzzy LSTM and Fuzzy CNN?

References

Alawad, H., An, M., & Kaewunruen, S. (2020). Utilizing an adaptive neuro-fuzzy inference system (ANFIS) for overcrowding level risk assessment in railway stations. *Applied Science, 10*, 5156.

Bakhtavar, E., Valipour, M., Yousefi, S., Sadiq, R., & Hewage, K. (2021). Fuzzy cognitive maps in systems risk analysis: A comprehensive review. *Complex & Intelligent Systems, 7*, 621–637.

Bao, R., Zhou, Y., & Jiang, W. (2022). FL-CNN-LSTM: Indoor Air quality prediction using fuzzy logic and CNN-LSTM model. *2nd International Conference on Electrical Engineering and Control Science (IC2ECS)*, 986–989.

Chopra, S., Dhiman, G., Sharma, A., Shabaz, M., Shukla, P., & Arora, M. (2021). Taxonomy of adaptive neuro-fuzzy inference system in modern engineering sciences. *Computational Intelligence and Neuroscience*, 2021, Article ID 6455592.

Gaur, S., Raju, K. S., Kumar, D. N., & Graillot, D. (2015). Multiobjective fuzzy optimization for sustainable groundwater management using particle swarm optimization and analytic element method. *Hydrologic Processes, 29*, 4175–4187.

Hsu, M. J., Chien, Y. H., Wang, W. Y., & Hsu, C. C. (2020). A convolutional fuzzy neural network architecture for object classification with small training database. *International Journal of Fuzzy Systems, 22*, 1–10.

Karaboga, D., & Kaya, E. (2019). Adaptive network-based fuzzy inference system (ANFIS) training approaches: A comprehensive survey. *Artificial Intelligence Review, 52*, 2263–2293.

Khanzadi, M., Nasirzadeh, F., & Dashti, M. S. (2018). Fuzzy cognitive map approach to analyze causes of change orders in construction projects. *Journal of Construction Engineering and Management, 144*, 04017111.

Langeroudi, M. K., Yamaghani, M. R., & Khodaparast, S. (2022). FD-LSTM: A fuzzy LSTM model for chaotic time-series prediction. *IEEE Intelligent Systems, 37*, 70–78.

Larrea, P. P., Zapata-Ríos, X., & Parra, L. C. (2021). Application of neural network models and ANFIS for water level forecasting of the salve faccha dam in the andean zone in northern ecuador. *Water, 13*, 2011.

Lence, B. J., Moosavian, N., & Daliri, H. (2017). Fuzzy programming approach for multiobjective optimization of water distribution systems. *Journal of Water Resources Planning and Management, 143*, 04017020.

Li, R., Hu, Y., & Liang, Q. (2020). T2F-LSTM method for long-term traffic volume prediction. *IEEE Transactions on Fuzzy Systems, 28*, 3256–3264.

Lin, C. J., & Jhang, J. Y. (2022). Intelligent traffic-monitoring system based on YOLO and convolutional fuzzy neural networks. *IEEE Access, 10*, 14120–14133.

Loucks, D. P., & Beek, E. V. (2017). *Water resource systems planning and management: An introduction to methods, models and applications*. Springer.

Morankar, D. V., Raju, K. S., & Kumar, D. N. (2013). Integrated sustainable irrigation planning with multiobjective fuzzy optimization approach. *Water Resources Management, 27*, 3981–4004.

Morankar, D. V., Raju, K. S., Vasan, A., & Vardhan, L. A. (2016). Fuzzy multiobjective irrigation planning using particle swarm optimization. *Journal of Water Resources Planning and Management. ASCE, 142*, 05016004.

Papageorgiou, E. I., & Salmeron, J. L. (2013). A review of fuzzy cognitive maps research during the last decade. *IEEE Transactions on Fuzzy Systems, 21*, 66–79.

Rao, S. S. (2013). *Engineering optimization: Theory and practice*, New Age International Publisher.

Reddy, M. J., & Kumar, D. N. (2020). Evolutionary algorithms, swarm intelligence methods and their applications in water resources engineering: A state-of-the-art review. *H2Open Journal*, 3, 135–188.

Ross, T. J. (2021). *Fuzzy sets and fuzzy logic with engineering applications*. Wiley.

Sada, S. O., & Ikpeseni, S. C. (2021). Evaluation of ANN and ANFIS modeling ability in the prediction of AISI 1050 steel machining performance. *Heliyon, 7*, e06136.

Sasikumar, K., & Mujumdar, P. P. (1998). Fuzzy optimization model for water quality management of a river system. *Journal of Water Resources Planning and Management, ASCE, 124*, 79–88.

Shruti, K., & Deka, P. C. (2020). A basic review of fuzzy logic applications in hydrology and water resources engineering. *Applied Water Science, 10*, 191.

Takagi, T., & Sugeno, M. (1985). Fuzzy identification of systems and its applications to modeling and control. *IEEE Transactions System Man Cybernetics, 15*, 116–132.

Vasan, A., Raju, K. S., & Pankaj, B. S. (2022). Fuzzy optimization based water distribution network design using self-adaptive cuckoo search algorithm. *Water Supply*, 22, 3178–3194 [open access paper under CC BY-NC-ND 4.0 License]

Vogeti, R. K., Jauhari, R., Mishra, B. R., Raju, K. S., & Kumar, D. N. (2024). Deep learning algorithms and their fuzzy extensions for streamflow prediction in climate change framework. *Journal of Water and Climate Change*, 15, 832–848 [open access paper under CC BY-NC-ND 4.0 License]

Zimmermann, H. J. (1991). *Fuzzy set theory and its applications*. Kluwer Academic.

Suggested Further Reading

Ang, Y. K., Talei, A., Zahidi, I., & Rashidi, A. (2023). Past, present, and future of using neuro-fuzzy systems for hydrological modeling and forecasting. *Hydrology, 10*, 36.

Apostolopoulos, I. D., & Groumpos, P. P. (2023). Fuzzy cognitive maps: Their role in explainable artificial intelligence. *Applied Sciences, 13*, 3412.

Frequently Asked Questions-ANFIS in the Fuzzy Logic Toolbox (http://www.cs.nthu.edu.tw/~jang/anfisfaq.htm, Accessed June 22, 2023.

Gu, X., Han, J., Shen, Q., & Angelov, P. P. (2023). Autonomous learning for fuzzy systems: A review. *Artificial Intelligence Review, 56*, 7549–7595.

Hajji, S., Yahyaoui, N., Bousnina, S., Brahim, F. B., Allouche, N., Faiedh, H., Bouri, S., Hachicha, W., & Aljuaid, A. M. (2021). Using a mamdani fuzzy inference system model (MFISM) for ranking groundwater quality in an agri-environmental context: Case of the hammamet-nabeul shallow aquifer (Tunisia). *Water, 13*, 2507.

Ivanova, M., Petkova, P., & Petkov, N. (2021). Machine learning and fuzzy logic in electronics: Applying intelligence in practice. *Electronics, 10*, 2878.

Karimi, N., Feylizadeh, M. R., Govindan, K., & Bagherpour, M. (2022). Fuzzy multiobjective programming: A systematic literature review. *Expert Systems with Applications, 196*, 116663.

Mittal, K., Jain, A., Vaisla, K. S., Castillo, O., & Kacprzyk, J. (2020). A comprehensive review on type 2 fuzzy logic applications: Past, present and future. *Engineering Applications of Artificial Intelligence, 95*, 103916.

Morankar, D. V. (2014). *Fuzzy-based Approach for Integrated Planning and Performance Evaluation of an Irrigation System*, Ph.D. Thesis, BITS Pialni.

Nair, A., Reckien, D., & van Maarseveen, M. F. A. M. (2019). A generalized fuzzy cognitive mapping approach for modelling complex systems. *Applied Soft Computing, 84*, 105754.

Poczeta, K., Papageorgiou, E. I., & Gerogiannis, V. C. (2020). Fuzzy cognitive maps optimization for decision making and prediction. *Mathematics, 8*, 2059.

Samavat, T., Nazari, M., Ghalehnoie, M., Nasab, M. A., Zand, M., Sanjeevikumar, P., & Khan, B. (2023). A comparative analysis of the mamdani and sugeno fuzzy inference systems for MPPT of an Islanded PV system. *International Journal of Energy Research, 1*, 7676113.

Szafranko, E., Srokosz, P. E., Jurczak, M., & Śmieja, M. (2022). Application of ANFIS in the preparation of expert opinions and evaluation of building design variants in the context of processing large amounts of data. *Automation in Construction, 133*, 104045.

Talpur, N., Abdulkadir, S. J., Alhussian, H., Hasan, M. H., Aziz, N., & Bamhdi, A. (2023). Deep neuro-fuzzy system application trends, challenges, and future perspectives: A systematic survey. *Artificial Intelligence Review, 56*, 865–913.

Varshney, A. K., & Torra, V. (2023). Literature review of the recent trends and applications in various fuzzy rule-based systems. *International Journal of Fuzzy Systems, 25*, 2163–2186.

Chapter 6
Emerging Research Areas

6.1 Introduction

The chapter discusses advanced topics, such as Blockchain, recent ML techniques, Evolutionary Algorithms (EA), AI Tools, the Internet of Things (IoT), Big Data, Decision Support Systems (DSS), Taguchi Design of Experiments, data augmentation, and Cross-Validation. Related information is as follows:

6.2 Blockchain

Water is an indispensable commodity of life, and its conservation is necessary. In this regard, scientific allocation of available water resources for drinking, farming, and industrial purposes is required for efficient utilization. One of the most promising approaches in this context is a Blockchain-based decentralized system that enables peer-to-peer trading of tokenized water (Li et al., 2022a). This section discusses the architecture of Blockchain, which comprises the application, consensus, network, and data layers (Fig. 6.1) in the context of water resources.

6.2.1 Architecture of Blockchain

Application layer comprises three categories of individuals, i.e., users, prosumers (a prosumer is a person who produces as well as consumes products), and administrators can access this layer. The following are the functionalities:

- Trading request: It facilitates a user to request for purchasing, and other users can negotiate prices for selling their water currency (1 water currency = X Rupee

K. Srinivasa Raju and D. Nagesh Kumar, *Artificial Intelligence and Machine Learning Techniques in Engineering and Management*,
https://doi.org/10.1007/978-981-96-2621-2_6

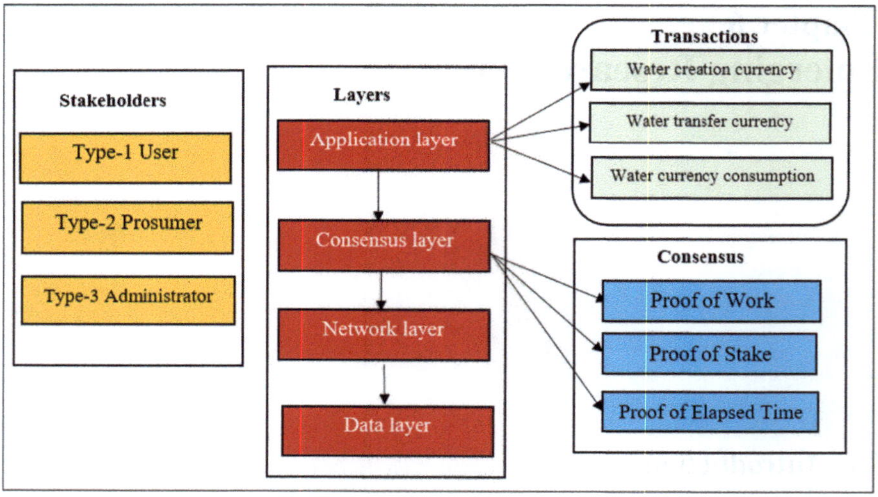

Fig. 6.1 Typical components of Blockchain

= 1 Litre (L) of permitted water usage). In this section, the currency and water currency are used synonymously.

- Transaction request: It is a portal for users to request transactions through the Blockchain.
- Block Explorer: It is a front end to view Blockchain data and all the transactions stored.
- Wallet: It helps users store their credentials (public or private key pairs) for their account transactions.
- User list: It is a front end where users can see a list of other active users and their water currency balance.
- User register: It is a portal for registering new users.
- Contract payment is the amount the user pays to specific contracts, i.e., public, private, or community, to receive water to their respective connections.
- Transaction validation: Every Blockchain protocol has a predefined set of transactions that will be considered valid. In general, these are SenderKey, ReceiverKey, and Amount.

Different access controls exist for various individuals, as discussed in Table 6.1.

The water currency Blockchain will allow three kinds of transactions to be written in the Blockchain, which are described as follows:

Water currency creation: This transaction enables registering when a sufficient amount of new water is available. They can only be considered valid and processed when data obtained from sensors support the transaction.

Water currency transfer: It facilitates the transfer of water currency from one user to another.

Table 6.1 Information about various stakeholders' access to different modules

Functionalities / Stakeholders	Trading request	Transaction request	Block explorer	Wallet	User list	User register	Contract payment	Transaction validation
Administrator	Yes	Yes	Yes	Yes	Yes	Yes	Yes	Yes
Prosumer	Yes	Yes	Yes	Yes	Yes	No	Yes	Yes
User	Yes	Yes	Yes	Yes	Yes	No	No	No

Water currency consumption: It indicates the water consumed by households, industry, farming, or other purposes. Water suppliers or local authorities and their respective smart contracts perform these kinds of transactions to register the amount of water consumed in the Blockchain.

The receiver field indicates where the water is consumed so that the water supplier can release the water to that connection after noticing that transaction. After that, the corresponding supplier for Meter_ID will release X litres of water to that connection. Any complex implementation can be performed using these three elementary transactions. Now, considering the balance of a user-type node in the system, unspent transaction output will be

$$\text{Balance} = \Sigma \text{ Currency bought} - \Sigma \text{ Currency sold} - \Sigma \text{ Currency consumed}$$

For the Prosumer or Administrator, it will be

$$\begin{aligned}\text{Balance} =& \Sigma \text{ Currency bought} - \Sigma \text{ Currency sold} + \Sigma \text{ Currency produced}\\ &- \Sigma \text{ Currency consumed.}\end{aligned}$$

The consensus layer determines the type of methods, like Proof of Work (PoW), Proof of Stake (PoS), and Proof of Elapsed Time (PoET)) used by all the nodes to accept a single Blockchain state all across the network and are explained as follows:

PoW: A group of nodes called miners takes a bunch of transactions from the pool and then forms a block. They must solve a computational hashing puzzle to add their block to the Blockchain network by competing with other miners. The miner unravelling the puzzle first will add their block to the chain, and everyone will accept that block. Consequently, they will be rewarded for doing so by the Blockchain protocol. However, after completion of the process, when a new block gets introduced to the Blockchain, all the miners need to start from scratch to find the appropriate Nonce (or, it can be said that, answer to the new puzzle as the last hash changes to the hash of the recently added block header). This results in the generation of 'Orphan Blocks,' the blocks generated by miners who cannot win the contest, which means that only one miner's work becomes useful, whereas others get wasted. In addition, the difficulty level of that computational puzzle is decided by the network's difficulty, which keeps increasing over time, making mining more and more challenging. Transactions related to Blockchain will be even higher in the water management context, so PoW is not a feasible solution.

PoS: One node is arbitrarily selected to add to the block in the Blockchain. Still, the chance of getting chosen is proportional to the amount of stake the node holds in the Blockchain token. It means that any stakeholder owns S out of 100 tokens in the network. They have an S% chance of signing the next block. So, the coins behave as collateral, and when a participant or node is selected to validate a transaction, they win a reward. Some variants of this algorithm work similarly, such as Proof of Coinage. This consensus allows a node to validate and add a block based on its share in the Blockchain network of tokenized assets. However, this approach cannot be followed for water management as it has few limitations. Firstly, it can lead to the scenario where water-rich regions or communities will have more control over the network than those with less, which is unacceptable in the case of water rights distribution. As a note, PoS and PoW are reward-driven consensus. However, no such reward is possible in the Blockchain network of tokenized water. Hence, such consensus cannot be applied to water management using Blockchain.

PoET is a Blockchain consensus algorithm preventing high resource utilization. Selecting the next network participant to add to the block is ensured randomly under the Trusted Execution Environment (TEE). Thus, there is no waste of computational resources or dependency on the stake of the participant to establish byzantine fault tolerance. TEE randomly generates a waiting time for each node, and the first node completing its waiting time will be allowed to be added to the network block. Here, the fairness of the randomness of the algorithm is ensured by a protected hardware environment. However, it also has a few limitations. First, it does not support openness as much as PoW or PoS because certification is required to join a PoET network. The second limitation is it does not reward the node for adding a block to the chain. So, the private participant has no incentive to make their computational resource available to the network.

The network layer takes care of communication between nodes of the Blockchain system. It will use the internet and can be quickly established using frameworks like Hyperledger.

The data layer defines information that will be stored inside the Blockchain. For example, if the block capacity is 4, four transactions per block can be implemented.

Some of the benefits of a Blockchain are:

- It is an immutable ledger that cannot be altered and tampered with. This aspect is ensured by cryptography, which links the small storage units called blocks. This linkage depends on the data stored in the block. Any attempt to alter the data will break the link or chain, making tampering evident. It can build systems with the inclusion of trust. Any stakeholder can see the transactions and transfers related to water from anywhere. Water thefts will be traceable, at least from well-known water resources that are linked with the system through sensors. An immutable ledger and the suitable facility to explore block data will allow every end user to make the right decision in a peer-to-peer exchange so that no one unfairly benefits from access to more information.

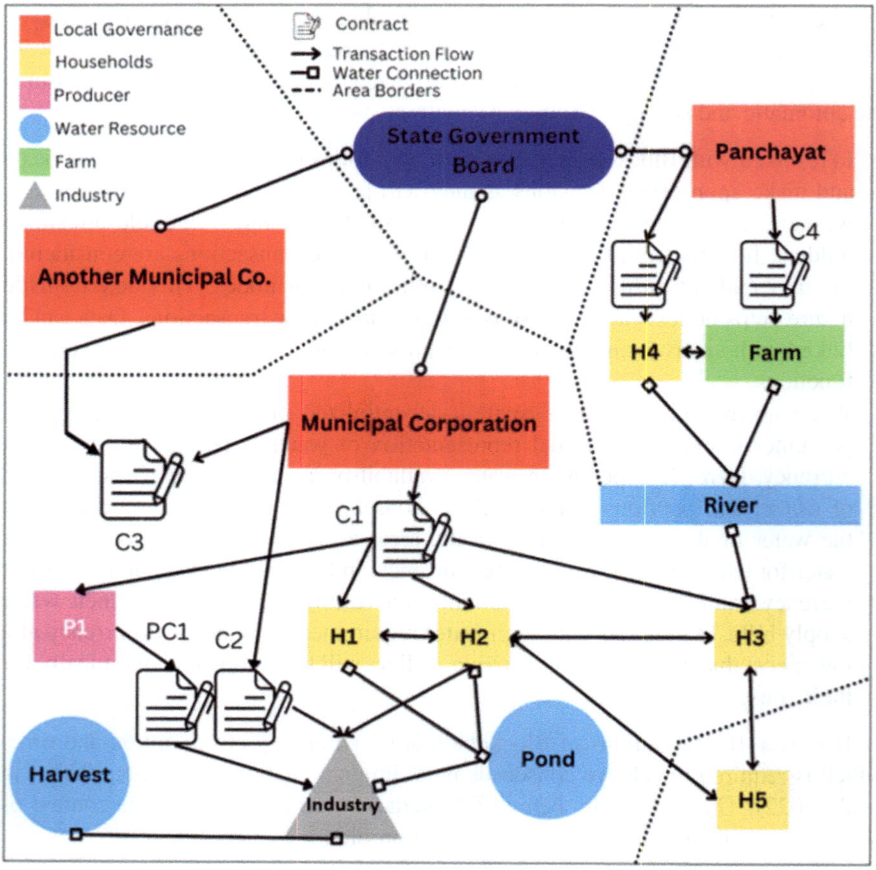

Fig. 6.2 Water management ecosystem

- Document icon pen represents the contracts (C1, C2, C3, C4) and prosumer contract (PC1).

Now understand the functioning of the system step by step, which is as follows (refer to Fig. 6.2):

1. First, the state authority is the administrator-level authority and is solely responsible for adding all the other entities in the system, including prosumers. Every prosumer-type authority has a water reservoir of its own. For example, the municipal corporation has a pond and a river shared with the panchayat nearby. In the case of shared water resources, a predefined agreement on the water distribution of that resource should exist. P1 is a prosumer and has a water harvesting facility.
2. When water intake is observed via IoT devices, the water currency creation transaction is registered by the respective authority having that resource or having a share in that resource.

SenderKey: ProducerKey, ReceiverKey: ProducerKey, Amount: Z

3. For every entity with water currency and being a prosumer type, the user releases smart contracts for different users through which they can buy currency by paying them in ₹. Here, the municipal corporation released contract 1 (C1) for household usage and prosumer P1, C2 for industrial use, and C3 for cooperation with another municipal corporation. Panchayat released contract C4 for farming usage. P1 also has a water transport facility and reached a contract agreement with industry users, creating a prosumer contract (PC1).
4. Terms in the smart contract can be different depending on the policy of the local authority. Various local authorities deploy smart contracts (like C1, C2, and C3 by the municipal corporation, C4 by the panchayat, and PC1 by prosumer P1) to structure their rates and usage regulations. They can be defined in their respective contracts.

For ease of understanding of the reader, C1 can be as follows for demonstration purposes,

- In C1, users will be charged ₹ X1 per 1 currency for water usage of 0 to 20 L, a day per person in the house.
- ₹ X2 per 1 currency will be charged for water usage of 20 to 50 L a day per person in house.
- ₹ X3 per 1 currency for water usage of more than 50 L a day per person in the house.

5. Now, every user is mapped by the water supplier (that can either be the local authority or any producer) to the contract they need to use to purchase water currency from them. For example, Households H1, H2, and H3 use contract C1. Although there can be a case where the contract will be the same, water supply might come from different sources depending on convenience. H1 and H2 receive water from the pond, but H3 gets water from the river. Also, contracts carry the data of users who are allowed to access it.
6. In this ecosystem, industry users have two contracts available, C2 and PC1, and they can choose to utilize any, depending on the terms and conditions, which can maximize their monetary benefits.
7. Having a transport network, municipal corporations can also establish more contracts like C3 with other authorities. This is how water can be directed from abundance to water-scarce areas.
8. Now, if a user H1 wants Y litres of water and according to their state of usage, water costs him ₹ X per currency or litre.

    - They will buy Y water currency from the C1 contract by paying ₹ X × Y. Then, the municipal corporation will initiate the following transaction.

SenderKey: MunicipalKey, ReceiverKey: H1Key, Amount: Y

- Then, H1 can spend that currency to get the water by performing water usage transactions.

SenderKey: H1Key, ReceiverKey: H1_METER_ID, Amount: Y

9. Users can also perform peer-to-peer water transfers. If H1 has reached a slot of ₹ X3 per litre, and user H2 has got into the X2 slab, where X3 > X2, then
   - User H1 will buy water currency from H2 at some cost if X2 < cost < X3.
   - If H1 buys W litres of water from H2 at some decided cost, they pay W × cost to H2, and H2 initiates the following water currency transfer transaction.

SenderKey: H2Key, ReceiverKey: H1Key, Amount: W

- Similar transactions can be performed by other households (like H3).
- Even if H5 belongs to other local governing authority areas, peer-to-peer transactions between H1, H2, H3, and H5 can also be enabled.

Ultimately, this can lead to more cooperation contracts between local authorities to ensure better water distribution. Further, this system will reward users economically and encourage less water usage and conservation. The proposed decentralized system is expected to improve water management, which means that the new system should be cost-effective for consumers and equitable water distribution among the people.

Numerical problem 6.1. Table 6.2 presents a hypothetical water distribution system's monthly water consumption (column 2). Compare the centralized system (Water bill @ ₹ 200 for unlimited consumption) and Blockchain, i.e., decentralized system (refer to Table 6.3). Table 6.3 presents a metered water supply with the new charges. The new tariff was to reduce the cost for families using less than 11,000 L of water only to ₹ 55 and charge more to households with more water usage (depending on the slabs). Discuss critically related aspects.

Solution:

Centralized System

Water bills (based on consumption) were presented in column 3 of Table 6.2 for each household. Here, the average household is charged ₹ 225.90**,** and the average monthly water consumption is 28080 litres.

Blockchain

See how the Blockchain can improve the situation using peer-to-peer transactions. It works on tokenizing water into water currency and an open market. Below are the salient points:

(1) Household 1 will trade water currency equivalent to 1100 L of water, with Household 12 providing him the water currency for anything between ₹ 7 and 8 so that their usage falls below 25,000 L.
(2) Household 2 will trade water currency promising 5400 L of consumable water to Household 6 for anything between ₹ 7 and 9.
(3) Household 3 will trade 10,300 tokens to Household 8 between ₹ 8 and 9.
(4) Similarly, Household 7 sells 600 L of water currency to Household 10.
(5) Household 9 also sells 2400 L worth of water currency to Household 10.

Table 6.2 Details of household numbers, water consumed per month, and water bill

Household number (1)	Water consumed (litre per month) (2)	Water bill (₹) (3)
1	22,500	157.5
2	18,200	127.4
3	36,000	288
4	42,000	336
5	9900	49.5
6	55,400	498.6
7	14,100	84.6
8	60,300	542.7
9	12,600	75.6
10	28,000	224
11	0	0
12	26,100	208.8
13	54,500	490.5
14	13,700	82.2
15	27,900	223.2

Table 6.3 Details of cost and lower and upper bounds of water usage

Charges in ₹ per 1000 L	Consumption lower bound (litre per month)	Consumption upper bound (litre per month)
5	0	11,000
6	11,000	15,000
7	15,000	25,000
8	25,000	50,000
9	50,000	∞

(6) Household 11 will sell 4500 L of water rights to Household 13, 2700 L of water rights to Household 14, and 2900 L worth of water currency to Household 15.

See the impact of introducing only eight peer-to-peer interactions in the system (Table 6.4).

Figure 6.3 presents water usage for the month after trading water rights to optimize the slot-bound utilization. Also, the average water usage is the same, but the standard deviation becomes 15,425.91 L, which was 18,190.62 L in the centralized scenario. It means that households have made better use of their water rights. The distribution has become fairer.

Figure 6.4 presents related water bills. The average cost of a household just after a few peer-to-peer interactions has decreased to ₹ 206.33, which is ₹ 19.57 less than the previous, so 15 households have saved a combined ₹ 293.55. The water is not consumed equally, but the low-consumption households have generated monetary

Table 6.4 Details of house numbers and water usage per month

Household number (1)	Water usage in litres per month (2)	Water bill (₹) (3)
1	22,500 + 1100 = 23,600	165.2
2	18,200 + 5400 = 23,600	165.2
3	36,000 + 10,300 = 46,300	370.4
4	42,000	336
5	9900	49.5
6	55,400–5400 = 50,000	400
7	14,100 + 600 = 14,700	88.2
8	60,300–10,300 = 50,000	400
9	12,600 + 2400 = 15,000	90
10	28,000–600–2400 = 25,000	175
11	0 + 4500 + 2700 + 2900 = 10,100	50.5
12	26,100–1100 = 25,000	175
13	54,500–4500 = 50,000	400
14	13,700–2700 = 11,000	55
15	27,900–2900 = 25,000	175
Average water use	28,080	206.33

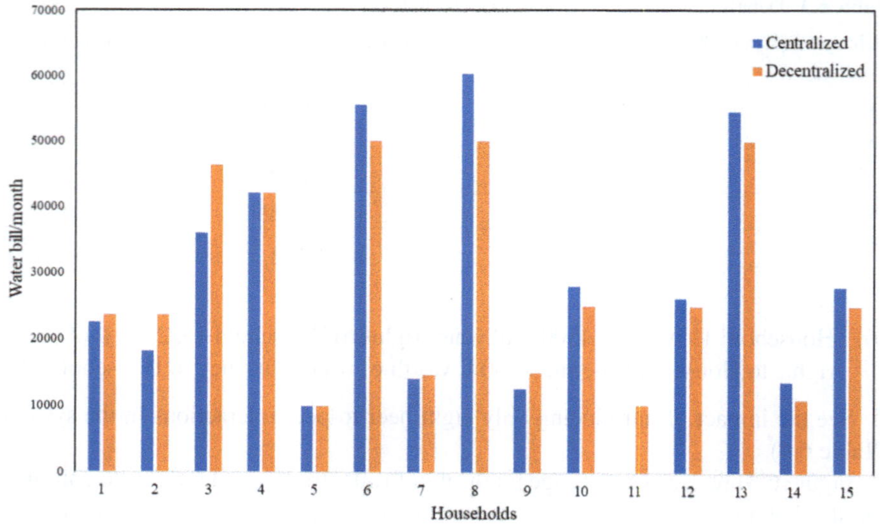

Fig. 6.3 Comparison of water usage in centralized and decentralized distribution systems

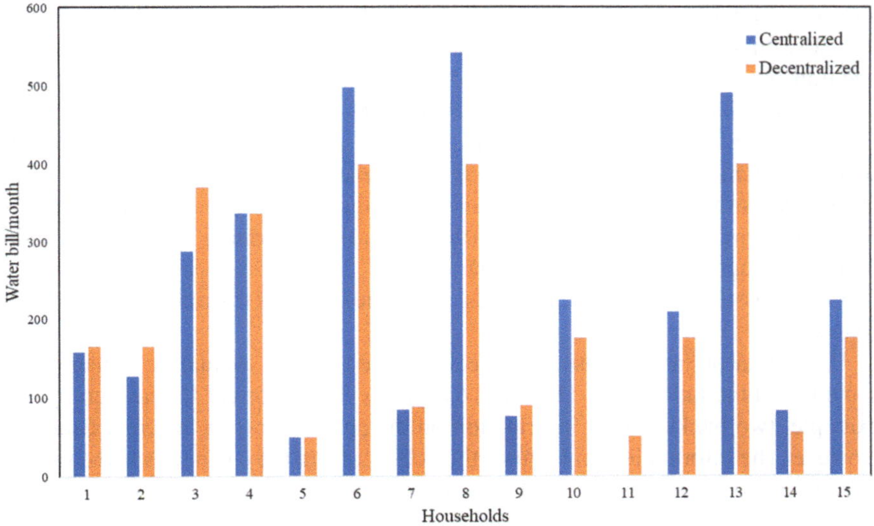

Fig. 6.4 Comparison of water bills in centralized and decentralized distribution systems

benefits from their water currency. For example, if Household 2 has traded water currency promising 5400 L of consumable water to Household 6 for ₹ 8.5 per kilo litre of water, it costs ₹ 7. It has generated 5.4 × (8.5 -7) = ₹ 8.1 of monetary benefit.

Also, the local authority spent ₹ 200 per household. However, it generates ₹ 206.33 (not ₹ 225.90), lower than before. It can be stated that a decentralized system is worthier than a centralized system.

The average cost decreases, distribution becomes fairer, and the low-consumption household generates revenue by trading their water rights. The effect of such an intangible system is that the households will be more aware of their water usage. Figure 6.4, related to monetary benefit, may seem small, but remember that peer-to-peer interaction is performed only between 15 households. Implementing the system on lakhs of users, including industrial and commercial, will significantly impact society.

6.3 Recent ML Techniques

6.3.1 Federated Learning

Almost all the ML algorithms work with a central learning philosophy, where data collected will be in one place and used for training. In Federated Learning (FL), a secured distributed learning framework, individual client (or user) edge devices train the given ML model parallelly (and locally) without moving data to a central computing facility. This is to respect data privacy and regulations within the individual entities. Also, decentralized computation, generalization, scalability, and transfer of only encrypted processed parameters make the collaborative process more reasonable, robust, and adaptive (Banabilah et al., 2022; Wen et al., 2022).

In a nutshell, individual devices train the models on the data available to them and send the parameters, such as weights, etc., in an encrypted format to a central server. It aggregates these comprehensively and transmits them back to the individual for further training until the model reaches the optimal state, terming it as a global model (Gupta & Gupta, 2023; Maroua, 2024).

FL is classified into three categories established on data distribution, which is as follows:

Horizontal or sample-based FL: Features used for the evaluation remained the same for all the individuals. However, datasets will be different for each entity. For example, two banks collect information on the customers (two different datasets). However, the information on features they collect from the customers is the same.

Vertical or feature-based FL: Different features and some overlapping datasets define this FL. For example, a person with the name X takes a bank loan [defined with features (p, q, r)] and invests it at another place [defined with features (s, t, u)]. When working on tax purposes, the name of person X is identified by the first set of features in one place and the second set of features in another place. Here, there is an overlap in the name. If required, these two sets of features can be combined to make it a complete dataset for person X.

Transferred FL: It is almost similar to vertical feature-based FL except for a small sample space.

Points to Be Noted

- There is no control over the quality of the data sent by individual devices.
- There is likely an imbalance of data while training by individual entities. It means there is no minimum threshold of datasets for training, and it will vary depending on the data available with individual entities.
- Datasets transmitted by each device are anticipated to be independent. Their distribution is expected to be identical. These requirements may not be feasible in most situations.
- Edge devices may be heterogeneous.

Some of the areas where potentiality exists are medical research, finance, or organizations.

6.3.2 *Neural Architecture Search*

Most of the architectures in deep learning are developed using trial and error approaches, which consumes considerable time. However, there is no guarantee that the architecture developed is optimal and suitable to the chosen problem (Chitty-Venkata et al., 2023).

Neural Architecture Search (NAS) is one of the algorithms that made inroads into this domain of finding logical architectures based on robust mathematical frameworks. NAS is employed by a number of researchers (Elsken et al., 2019; Poyser & Breckon, 2024). It works on three following principles:

Search space: Possible architectures that can be studied. Earlier knowledge of architecture may ease the search space.

Search strategy: Mechanism of identifying the best-performing architecture from search space. Some of the search strategies include EA, Random Search, and Bayesian Optimization.

Performance evaluation of search strategy: Estimating the efficacy of search strategy.

6.3.3 Miscellaneous Techniques

There are a number of ML techniques that have a lot of potential to be applied to real-world problems. A list of those techniques with related references is provided in Table 6.5 for the benefit of readers.

Table 6.5 Additional techniques falling under advanced aspects of ML techniques

Topic (in alphabetical order)	References
Autoencoders	Chen and Guo (2023), Li et al. (2023a), Berahmand et al. (2024), Qian et al. (2022)
Auto ML	Salehin et al. (2024), Baratchi et al. (2024), Singh and Joshi (2022), Vaccaro et al. (2021)
Capsule Networks	Patrick et al. (2022), Haq et al. (2023), Pawan and Rajan (2022), Mazzia et al. (2021)
Deep Q Networks	Jain et al. (2022), Huang (2020), Talaat (2022)
Explainable Artificial Intelligence:	Ali et al. (2023), Naser (2021), Tantithamthavorn and Jiarpakdee (2021), Ghosh et al. (2024), Love et al. (2023)
Generative Adversarial Network	Aggarwal et al. (2021), Gonog and Zhou (2019), Lee (2023), Nayak et al. (2024), Jozdani et al. (2022)
Graph Neural Networks	Khemani et al. (2024), Zhou et al. (2020), Corso et al. (2024), Besharatifard and Vafaee (2024), Sun et al. (2023)
Neural Network Pruning and Quantization	Liang et al. (2021), Zhang et al. (2022), Alqahtani et al. (2021), Cai et al. (2023a)
Neural Style Transfer	Singh et al. (2021), Cai et al. (2023b), Li et al. (2020a)

6.4 Evolutionary Algorithms

EAs are rapidly expanding their role in AI and allied fields, and most of them are motivated by the behaviour of living beings or nature. They are most flexible and can easily handle the challenges experienced by traditional algorithms, like complex constraints, local optima, and high dimensional non-linear problems. These situations are prevalent in real-world planning problems, where near-optimal solutions suffice. However, one bottleneck is that almost all algorithms are parameter-dependent. Sometimes, it is arduous to identify the precise values of these parameters that best fit the chosen problem. These algorithms are classified mainly into several categories. However, two salient categories are briefly discussed here.

Biologically inspired EA consist of a population comprising a number of individuals, each characterizing a search point in the feasible solution space. The workflow starts with random initialization of population, selection, recombination, and mutation, and the process continues through several generations. The fitness of all the individuals is estimated. Individuals with worthier fitness are combined to create new individuals who may have better fitness than the previous generation. This activity continued until there was no change in fitness value in the successive generations (Reddy & Kumar, 2020). A sample structure is shown in Fig. 6.5. Two major factors that hinder the successful evolution process are selective pressure and population diversity. Reddy & Kumar (2020) provided detailed information about handling these challenges.

Behaviourally Inspired Swarm Intelligence (SI)-based algorithms are established on socio-cognition, which can be applied to unravel different optimization tasks. They are also population-based and similar to EA. However, mutation and recombination are not part of this scheme. The members of a swarm work without any guidance and have stochastic behaviour. They utilize resources competently through collective group intelligence. A significant characteristic is self-organization, which facilitates the evolution of global-level responses employing local-level interactions. The system is randomly initialized with a population of individuals. These are then evolved over a number of generations by capturing the insect's social behaviour to determine the optimal (Reddy & Kumar, 2020). Several excellent papers on meta-heuristics and related topics are available. Representatives are presented in Table 6.6 for the benefit of readers.

Several benchmarking functions are available for utilization in algorithms mentioned in Table 6.6 to evaluate their performance for single-objective and multi-objective optimization problems (Hellwig & Beyer, 2019; Piotrowski et al., 2023; Volz et al., 2023). Later, suitable among these can be employed to unravel real-world challenging problems, which will increase policymakers' confidence in possible implementation.

Representative test functions employed in the single objective optimization category are Ackley's, Bohachevsky, Booth's, Bukin N.6, Colville, Drop wave, Easom, Eggholder, Goldstein-Prince, Griewank, Holder Table, Matyas Function, Michalewicz, Rastrigin, Rosenbrock (Banana), Schaffer N.2, Schwefel, Shekel,

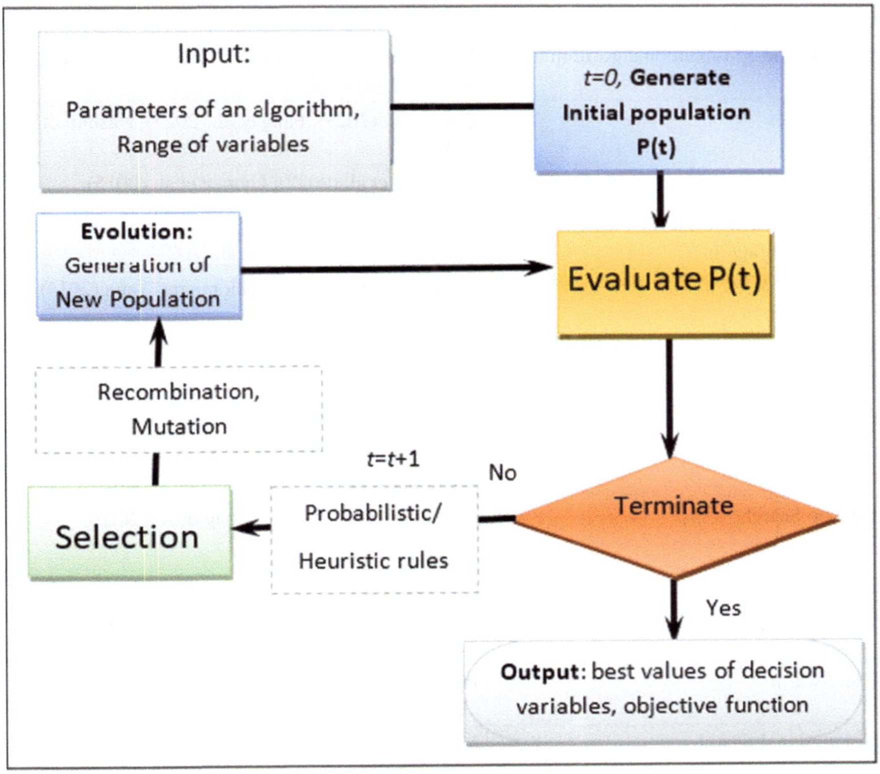

Fig. 6.5 The basic structure of an EA (Adapted from Reddy & Kumar, 2020 under CC BY- 4.0 License)

Shubert, Six Hump, Camel Back, Sphere (Molga & Smutnicki, 2005; Surjanovic & Bingham, 2013). In the case of multiobjective optimization, the ZDT test suite, which comprises six different test problems, is employed by a number of researchers (Zitzler et al., 2000). These problems consist of two objectives, constraints and bounds.

Note that these are representative test functions. Some others also exist. Readers are encouraged to study those for a better understanding.

6.5 Large Language Model (LLM)-Based Generative AI

LLM-based generative AI has the potential to simulate human-like dialogues and interactive, high-quality, and much more (https://www.elastic.co/what-is/large-language-models). Several related AI tools exist for the mentioned functionality and are presented in Table A.1 of Appendix A.

Table 6.6 Representative reference(s) where the EAs were discussed

Name of the optimization algorithm topic	Representative reference(s) where the technique is discussed in detail
Archimedes	Dhal et al. (2023), Fang et al. (2023), Hashim et al. (2021)
Artificial Algae	Turkoglu et al. (2022), Uymaz et al. (2015)
Artificial Bee Colony	Xiao et al. (2023), Zhao et al. (2022)
Bacterial Foraging	Chen et al. (2020), Guo et al. (2021)
Bat	Gagnon et al. (2020), Gandomi and Yang (2014), Shehab et al. (2023)
Biogeography	Sang et al. (2021), Wei et al. (2022)
Black Hole	Abualigah et al. (2022a), Deeb et al. (2022)
Chicken Swarm	Wang et al. (2023a), Liang et al. (2023), Zhang et al. (2023), Zouache et al. (2019)
Crow Search	Askarzadeh (2016), Hussien et al. (2020)
Cuckoo Search	Mohamad et al. (2014), Xiong et al. (2023)
Elephant Clan	Jafari et al. (2021)
Elephant Herding	Li et al. (2020b)
Equilibrium	Elmanakhly et al. (2021), Faramarzi et al. (2020), Yang et al. (2022a)
Fish Swarm	Pourpanah et al. (2023), Tan and Mohammad-Saleh (2020)
Gravitational Search	Mittal et al. (2021), Rashedi et al. (2009), Yang et al. (2022b)
Grey Wolf	Mirjalili et al. (2014), Pan et al. (2021), Wang and Li (2019)
Harmony Search	Kim (2016), Wang et al. (2023b)
Henry Gas Solubility	Hashim et al. (2019), Li et al. (2022b), Mohammadi et al. (2022)
Honey Badger	Hashim et al. (2022)
Imperialist Competitive	Abdollahi et al. (2013), Bernal et al. (2017), Hosseini and Al Khaled (2014)
Indicator	García et al. (2021), Yuan et al. (2022), Li et al. (2023b)
Jellyfish Search	Chou and Molla (2022), Chou and Truong (2021), Manita and Zermani (2021)
Krill Herd	Bolaji et al. (2016), Wang et al. (2019), Gandomi and Alavi (2012)
Learner performance-based behaviour	Rahman and Rashid (2021)
Meta-Heuristic	Abdel-Basset et al. (2018), Alorf (2023), Khanduja and Bhushan (2020), Rajwar et al. (2023), Rajalakshmi and Kanmani (2022), Velasco et al. (2024), Wong and Ming (2019)

(continued)

Table 6.6 (continued)

Name of the optimization algorithm topic	Representative reference(s) where the technique is discussed in detail
Moth Flame	Mirjalili (2015), Nadimi-Shahraki et al. (2023a), Sahoo et al. (2023), Shehab et al. (2020)
Mountain Gazelle	Abdollahzadeh et al. (2022)
Mountaineering Team	Faridmehr et al. (2023)
Multiverse	Benmessahel et al. (2020), Mirjalili et al. (2016)
Red Fox	Połap & Woźniak (2021)
Reptile Search	Abualigah et al. (2022b), Khan et al. (2023), Sasmal et al. (2024)
Salp Swarm	Duan et al. (2021), Hegazy et al. (2020), Mirjalili et al. (2017)
Sand Cat swarm	Kiani et al. (2023), Wu et al. (2022), Seyyedabbasi & Kiani (2023)
Shuffled frog leaping	Maaroof et al. (2022), Zhao et al. (2024)
Social Spider	Feng et al. (2022), Yu and Vok (2015), Zhao et al. (2017)
Spider Monkey	Agrawal et al. (2023)
Swarm Intelligence	Brezočnik et al. (2018), Tang et al. (2021), Figueiredo et al. (2019)
Symbiotic Organisms Search	Ezugwu & Prayogo (2019), Cheng & Prayogo (2014), Gharehchopogh et al. (2020)
Teaching–Learning	Gómez Díaz et al. (2022), Xu et al. (2022)
Teamwork	Dehghani & Trojovský (2021)
Walrus	Han et al. (2024)
Whale	Mirjalili & Lewis (2016), Nadimi-Shahraki et al. (2023b), Rana et al. (2020)
Wild Horse	Naruei & Keynia (2022), Zheng et al. (2022)

One of the most prominent tools in this category, the Chat Generative Pre-Trained Transformer (ChatGPT), is discussed in detail for the benefit of the readers. It is established on Deep Neural Network (DNN) architecture with unsupervised pre-training initially and later supervised fine-tuning.

The latest version in this series, GPT 4.0, provides improved Natural Language Processing (NLP) abilities, handling large datasets, more extended consecutive outputs, and contextual inferencing, resulting in an engaging conversation. Its flexibility to provide relevant information in any domain, including coding tasks and correcting program errors, makes ChatGPT unique (Ray, 2023). Dempere et al. (2023) studied the effect of ChatGPT on higher education, Foroumandi et al. (2023) on hydrology and earth sciences, Huang & Tan (2023) on scientific communication and Nikolic et al. (2023) on assessment of engineering education.

A comparison of ChatGPT and Google Bard is made by McGowan et al. (2023) for the psychiatry literature search context and Cheong et al. (2023) for patient education material for obstructive sleep apnoea. Furthermore, Agarwal et al. (2023) applied ChatGPT, Bing, and Bard to develop reasoning-based questions in the domain of Medical Physiology, and Lim et al. (2023) in the context of trauma nerve laceration patients. The strengths and limitations of ChatGPT are discussed in detail.

Strengths

1. It saves considerable time and improves productivity. Relevant lengthy keywords with a focussed domain help explore ChatGPT to its full potential. Otherwise, it may give some output that may not be related to your query.
2. Diverse content can likely be generated with similar keywords, which may provide a broader view to the user.

Limitations

1. Limited exposure of ChatGPT for a specific query for which it was not trained may result in inaccurate or incomplete output.
2. The resulting output may not be authentic, necessitating a further investigation before utilization.
3. There is a risk of plagiarism for generated content. It is always advisable to verify the generated content with plagiarism-related software.
4. A graphical visualization facility is not available at present. This means that considerable effort is required to understand the intricacies of the results and possible inferences.
5. Memory may become exhausted during the processing of a large amount of conversations and other related tasks.

In summary, disrupting users' creative thinking processes, skill sets, and expression abilities is a significant concern. However, limiting AI tools is not the solution, and a trade-off is necessary (Nah et al., 2023; Ray et al., 2024; Roumeliotis & Tselikas, 2023). The impact of these AI tools on society must also be studied holistically. Most of the strengths and limitations mentioned in ChatGPT are also applicable to some of the tools mentioned in Table A.1. Remarks in Table A.1 are based on the limited understanding of the tools by the authors of the book. Readers are strongly advised to verify thoroughly before working on any tool.

6.6 IoT, Big Data, and DSS

IoT is a network of objects established with sensors and devices and has connectivity capabilities. It has gained much prominence due to its strength to monitor data in real time. It empowers them to gather and swap data over the internet for further modelling applications. It also provides flexibility to make quick decisions and understand risks and vulnerabilities. This, in turn, helps assess proactive and predictive maintenance requirements, which minimize downtime. The architecture comprises four layers: sensing, network, data processing, and application. Relevant information is as follows (Jena, 2023):

1. The sensing layer is the first layer for gathering data from various sources. It consists of sensors and other measuring devices to collect information, such as the number of vehicles, temperature in a lake, pollution at a given location, etc.
2. The network layer connects with the sensing layer with wireless or wired communication protocols. Its primary function is to facilitate information sharing between devices in the IoT structure.
3. The data processing layer collects data from the devices using software and hardware components and processes it for further analysis. One example is using data in AI, EA, and other models where their outcomes provide insights to decision-makers.
4. The application layer is the user interface layer.

In IoT and similar data acquisition procedures, large volumes of data will be continuously generated from various sources, termed Big data (Kapliński et al., 2016). Big data analysis is helpful for process improvements, predicting long-term trends, and many more. However, the following challenges remain to utilize generated data to the fullest extent (Thayyib et al., 2023):

- Data quality: Available data are expected to be of good quality without inconsistencies and errors. This may not be possible sometimes due to multiple data sources. For example, land use information is retrieved from satellite images, water quality is acquired from sensors, and reservoir inflows are collected from conventional measuring devices. There must be efficient data integration and verifying mechanisms, which are expected to save the analyst's time. In this context, sensors and other data-acquiring devices are to be calibrated to create confidence in the measurement system.
- Data privacy: There is a likelihood that data may be leaked, stolen, or manipulated during the transmission stage, which may lead to erroneous outcomes from the modelling. Encryption and cyber security are some measures that can be explored to minimize the impact.
- Human resources: The requirement of skilled experts for big data analytics is a bottleneck. It can be overcome by training the human resources to be focused and sustainable.
- Data storage: Cost-effective processing of ever-growing data is a perennial challenge. One of the solutions may be to use cloud facilities.

As mentioned earlier, the data generated from the procedures are valuable inputs to DSS and assist policymakers in a complex, data-driven environment. Its adaptability to technology, data analysis, and models is an added advantage driven by the following components (Alamanos et al., 2021; Alshami et al., 2023; Gheibi et al., 2023).

- Models: They are crucial in decision-making. They will be helpful in forecasting outcomes for different possible scenarios expected to yield multi-faceted decisions.
- User interface: A window through which policymakers interact with the DSS. It consists of data visualizations in graphical format, dashboard, and query interfaces.
- Knowledge-based system: Domain-specific knowledge provided by the expert and resulting rules developed by the analyst guide the decision process.
- Decision process: It depends entirely on previous components for formulating logic that drives decisions.

However, in most DSSs, almost all the components mentioned here are involved in successful decisions. The involvement of experts is of paramount importance, as they can identify whether the decision was implemented and the effects of the decision are as expected after implementation. The feedback mechanism is an essential and inevitable task that facilitates continuous improvement of DSS over time.

6.7 Taguchi Design of Experiments

Taguchi's approach, a statistical method for the design of experiments, became prominent due to the advantage of identifying factors and levels among the available that contribute significantly to the outcome (Ginting & Tambunan, 2018; Kacker et al., 1991). This process also substantially enhances the efficacy of ML algorithms in the context of parameter tuning and selecting only relevant factors and levels (Li et al., 2024). It has two working principles: an orthogonal array and a signal-to-noise (S/N) ratio. Before moving forward, a brief introduction of terminology helps to understand the philosophy of the Taguchi approach.

Factor: Variable, feature, or parameters.

Response: Outcome.

Levels: Discrete values of factor(s) between lowest and highest values (for example, 20 and 100). Researchers can divide the range into any number of levels depending on the problem requirement and computational resources available. For example, levels can be 20, 40, 60, 80, 100 (representing five levels) and denoted respectively as 1, 2, 3, 4, 5.

Suppose a researcher is considering five factors and two levels. In that case, 2^5, i.e., 32 simulation runs, are required to thoroughly study the chosen problem (Table 6.7), which is termed a full factorial design. Sometimes, these are manageable for conducting simulation runs.

However, if there are six factors and four levels, the runs required is 4^6, i.e., 4096, which is impractical for performing the simulation. This means that with an increment in levels and factors, the number of runs will increase.

Table 6.7 Full factorial design

Run	Feature 1	Feature 2	Feature 3	Feature 4	Feature 5
1*	1	1	1	1	1
2	1	1	1	1	2
3	1	1	1	2	1
4*	1	1	1	2	2
5	1	1	2	1	1
6	1	1	2	1	2
7	1	1	2	2	1
8	1	1	2	2	2
9	1	2	1	1	1
10	1	2	1	1	2
11	1	2	1	2	1
12	1	2	1	2	2
13*	1	2	2	1	1
14	1	2	2	1	2
15	1	2	2	2	1
16*	1	2	2	2	2
17	2	1	1	1	1
18	2	1	1	1	2
19	2	1	1	2	1
20	2	1	1	2	2
21	2	1	2	1	1
22*	2	1	2	1	2
23*	2	1	2	2	1
24	2	1	2	2	2
25	2	2	1	1	1
26*	2	2	1	1	2
27*	2	2	1	2	1
28	2	2	1	2	2
29	2	2	2	1	1
30	2	2	2	1	2
31	2	2	2	2	1
32	2	2	2	2	2

* Part of an orthogonal array, L_8

In these types of situations, an orthogonal array, a downsized or fraction of full factorial design, plays a significant role. It is a two-dimensional matrix. Column denotes employed factors, whereas row denotes simulation runs with information of levels for the employed factors. For the problem of 2 features and 5 levels, an orthogonal array is L_8. It means eight simulation runs, which is a substitution for 32 runs. It reduces computational requirements and is scientifically validated (Cimbala, 2014; Fraley et al., 2024). (*) repesrents Orthogonal Array in Table 6.7.

The S/N ratio is a measure to understand the efficacy of each simulation run in the context of smaller being better (in case of cost), larger being better (in case of benefit), or nominal being the best. These situations are based on the chosen criteria.

In summary, the following are the steps.

1. Identify the relevant factors, number of levels (corresponding values for each factor), and orthogonal array.
2. Execute the simulation for the number of rows in the orthogonal array and analyze.
3. Computation of S/N ratio.
4. Identify key levels of factors and further analysis.

A detailed procedure of Taguchi with a demonstrative example is provided by Ginting & Tambunan (2018). Frey (1998) provided excellent information about orthogonal arrays.

You can find the number of research papers associated with the combined application of the Taguchi approach and ML. Representative applications where the Taguchi approach is part are optimization of the hybrid cooling array and LSTM (Li et al., 2024), Hydrogen separation and decision tree (DT), SVM and ensemble method (Chen et al., 2024), and Smart manufacturing systems and ML (Nejati et al., 2024).

6.8 Data Augmentation

It is the process of adding artificial data when available real-world data is too small (or not available) or too costly to acquire for training purposes of ML algorithms. It is a known fact that ML algorithms require large amounts of diverse data to bring meaningful inferences.

If added scientifically, there is likely a chance of improved generalization, minimization of over-fitting, and improved accuracy. One of the ways is to modify the original real-world data with minor changes. Another way to synthetically generate high-quality data is by using generative adversarial networks, variational autoencoders, deep neural networks, neural style transfers, or similar algorithms.

You can find a number of applications of data augmentation in healthcare (Garcea et al., 2023; Goceri, 2023), Finance (Ranjbaran et al., 2023), Water Quality (Mahlathi et al., 2022), Power Sector (Chen et al., 2021), and more. Representative data sources are available in Table A.2 of Appendix A.

6.9 Cross-Validation

Here, the model is executed on the number of sub-datasets (or folds). One of the folds is used for validation, whereas the remaining folds are for training. Iteratively, the fold used for validation changes, and accordingly, the remaining folds are employed for training (Madhuri et al., 2021). Accordingly, performance measures can be computed for each iteration, and the average performance measure is used as the basis to understand the model efficacy. It is also one of the approaches to make the model more generalized without over-fitting (it is a challenge in the ML model, where it performs well for training and is unsatisfactory on the testing data).

There are also a number of approaches in this category, such as Holdout validation, K-fold cross-validation, and more. In the case of K-fold, it aggregates results from various chosen training and testing datasets and shows an unbiased picture of the algorithm performance on a given dataset.

Tougui et al. (2021) employed cross-validation techniques for diagnostic applications, Kaliappan et al. (2023) for Early Detection of Intrauterine Fetal Demise, whereas Kee et al. (2023) for Smart and Lean Pick-and-Place Solution.

Revision Questions and Exercise Problems

6.1 What is the functionality of Blockchain?
6.2 What is the difference between a centralized and decentralized system?
6.3 What are smart contracts?
6.4 What are the features of immutability in Blockchain? How are these going to help or accelerate the decentralized process?
6.5 What are blocks and nodes in the water management ecosystem?
6.6 What may be the function of the stakeholders in the Blockchain process?
6.7 How does IoT help the Blockchain process?
6.8 What is tokenization in Blockchain?
6.9 What is water currency?
6.10 What are the benefits of Blockchain? Discuss critically.
6.11 What is the purpose of consensus algorithms? What are popular consensus algorithms?
6.12 How do PoW and PoS differ?
6.13 Who are miners?
6.14 What is the computational process of adding the block to the Blockchain?
6.15 What is meant by a puzzle in the Blockchain process?
6.16 What is meant by reward and hash?
6.17 What is Nonce?
6.18 What is PoET? Mention a few limitations of implementing PoET.
6.19 Discuss the architecture of Blockchain.
6.20 What is the difference between water currency creation, transfer, and consumption?
6.21 What is FL? What are the various steps in this modelling technique?
6.22 What is NAS? Explain mathematical philosophy.

The smart contract is a self-executed undertaking between the seller and buyer. These contracts run on the Blockchain network and can perform tasks previously done by a centralized authority. The decentralized character of Blockchain ensures the automatic and speedy execution of contracts without intermediaries.

- IoT, linked with Blockchain, can benefit authorities by allowing them to allocate and make appropriate decisions as they will provide precise data. Water usage, wastage, and quality can be monitored and made available to the relevant stakeholders. It starts from the first block, and if all the transactions are considered, the user ends up in the same state as other users or nodes. This deterministic nature helps nodes reach a consensus and approve new transactions. Data can be linked with smart contracts, making contracts more effective in their respective functions.
- Water needs tokenization to create a decentralized management system and a prosumer market. The virtual representation of water is represented by water currency. Here, X depends on water availability in that particular region, type of usage, and previous usage, such as household or industrial. Depending on the water intake and availability in the area, every household is granted some water for usage in the form of water currency. Industries can also purchase water currency from authorities or community harvesting facilities to pay their water supply bills. Individuals with high water requirements can buy this currency at a lower cost than the authorities provide. Sellers will be rewarded economically for their conservation efforts.

However, the limitations of Blockchain are high energy consumption and time, which is required to achieve consensus in its implementations (Sriyono, 2020; Xia et al., 2022). The philosophy behind Blockchain is demonstrated in the context of water management, followed by a numerical problem.

6.2.2 *Water Management Ecosystem*

Here, real-life scenarios are presented to understand how this system can be implemented and might work on the ground level (refer to Fig. 6.2). In Fig. 6.2,

- The entity in purple is the state government board, considered administrator (type 3 category).
- Entities in red are local governance, and pink is a prosumer (both are type 2 categories).
- Yellow, green, and grey are household [H], farm, and industry users, respectively (type 1 category).
- The blue elements in the diagram are the water bodies [pond, river, and harvest].
- Dotted lines define local area borders, squared or circular lines represent metered water connections, and arrows indicate transaction flow.

6.23 List some of the techniques that are in the advanced category of ML techniques.
6.24 List some of the techniques that are in the advanced category of optimization techniques.
6.25 What are the meta-heuristic optimization algorithms?
6.26 What are different swarm intelligence methods? On what mathematical basis were they developed?
6.27 What is meant by AI tools?
6.28 What is the purpose of ChatGPT?
6.29 What is IoT? How does it help to improve the data collection process?
6.30 How are IoT and Big Data related?
6.31 What is a DSS? How many components are part of it?
6.32 What are the advantages and disadvantages of a DSS?
6.33 What is Taguchi's design of experiments?
6.34 What is data augmentation? In what situations can it be used?
6.35 What is cross-validation?

Advance Review Questions

6.36 Discuss two case studies related to Blockchain.
6.37 Discuss the potential applications of Blockchain in engineering.
6.38 Do you think Blockchain is suitable for water resources in developing countries? If yes or no, justify the same.
6.39 How do you select the suitable algorithm for your domain of interest? What are the criteria you use for this purpose?
6.40 What is the meaning of self-adaptive? How does it work? Mention four case studies where a self-adaptive mechanism was employed.
6.41 Is there any possibility to relate ML and optimization algorithms? If yes, discuss it critically. If no, provide the reason why it is not possible.
6.42 Can you mention the name of any algorithm universally applicable to all situations?
6.43 Discuss four case studies related to IoT in your domain of interest.
6.44 Do you think DSS applies to AI in the present scenario? Critical discussion is highly suggested, preferably with examples.
6.45 Is there any possibility that DSS can be developed in a fuzzy context? If so, how they can function?
6.46 Are you encouraging AI tools to mimic human logic? Discuss the same in detail in the context of literature review, examination, and syllabus preparation.

References

Blockchain

Li, Y., Xie, J., Yang, J., Ren, J., & Zhai, N. (2022a). Application of Blockchain technology in water rights trading in the irrigation area under the internet-of-things environment. *Security and Communication Networks*, *2022*, Article ID 8700730.

Sriyono, E. (2020). Digitizing water management: Toward the innovative use of Blockchain technologies to address sustainability, *Cogent Engineering*, 7, Article no: 1769366.

Xia, W., Chen, X., & Chao, S. (2022). A Framework of Blockchain technology in intelligent water management. *Frontiers in Environmental Science*, 10–2022.

Recent Machine Learning Techniques

Autoencoders

Berahmand, K., Daneshfar, F., Salehi, E. S., Li, Y., & Xu, Y. (2024). Autoencoders and their applications in machine learning: A survey. *Artificial Intelligence Review, 57*, 28.

Chen, S., & Guo, W. (2023). Auto-encoders in deep learning—a review with new perspectives. *Mathematics, 11*, 1777.

Li, P., Pei, Y., & Li, J. (2023a). A comprehensive survey on design and application of autoencoder in deep learning. *Applied Soft Computing, 138*, 110176.

Qian, J., Song, Z., Yao, Y., Zhu, Z., & Zhang, X. (2022). A Review on autoencoder based representation learning for fault detection and diagnosis in industrial processes. *Chemometrics and Intelligent Laboratory Systems, 231*, 104711.

Auto ML

Baratchi, M., Wang, C., Limmer, S., van Rijn, J. N., Hoos, H., Bäck, T., & Olhofer, M. (2024). Automated machine learning: Past, present and future. *Artificial Intelligence Review, 57*, 122.

Salehin, I., Islam, S., Saha, P., Noman, S. M., Tuni, A., Hasan, M., & Baten, A. (2024). AutoML: A systematic review on automated machine learning with neural architecture search. *Journal of Information and Intelligence, 2*, 52–81.

Singh, V. K., & Joshi, K. (2022). Automated machine learning (AutoML): An overview of opportunities for application and research. *Journal of Information Technology Case and Application Research, 24*, 75–85.

Vaccaro, L., Sansonetti, G., & Micarelli, A. (2021). An empirical review of automated machine learning. *Computers, 10*, 11.

Capsule Networks

Haq, M. U., Sethi, M. A. J., & Rehman, A. U. (2023). Capsule network with its limitation, modification, and applications—a survey. *Machine Learning and Knowledge Extraction, 5*, 891–921.

Mazzia, V., Salvetti, F., & Chiaberge, M. (2021). Efficient-CapsNet: Capsule network with self-attention routing. *Scientific Reports, 11*, 14634.

Patrick, M. K., Adekoya, A. F., Mighty, A. A., & Edward, B. Y. (2022). Capsule networks—a survey. *Journal of King Saud University—Computer and Information Sciences, 34*, 1295–1310.

Pawan, S. J., & Rajan, J. (2022). Capsule networks for image classification: A review. *Neurocomputing, 509*, 102–120.

Deep Q-Networks

Huang, Y. (2020). Deep Q-networks. In H. Dong, Z. Ding, & S. Zhang (Eds.), *Deep Reinforcement Learning*, 135–160. Springer.

Jain, A., Mehrotra, A., Rewariya, A., & Kumar, S. (2022). A systematic study of deep Q-networks and its variations. 2022 2nd International Conference on Advance Computing and Innovative Technologies in Engineering (ICACITE), India, 2157–2162.

Talaat, F. M. (2022). Effective deep Q-networks (EDQN) strategy for resource allocation based on optimized reinforcement learning algorithm. *Multimedia Tools and Applications, 81*, 39945–39961.

Explainable Artificial Intelligence

Ali, S., Abuhmed, T., El-Sappagh, S., Muhammad, K., Alonso-Moral, J. M., Confalonieri, R., Guidotti, R., Ser, J. D., Díaz-Rodríguez, N., & Herrera, F. (2023). Explainable artificial intelligence (XAI): What we know and what is left to attain trustworthy artificial intelligence. *Information Fusion, 99*, 101805.

Ghosh, S., Kamal, M. S., Chowdhury, L., Neogi, B., Dey, N., & Sherratt, R. S. (2024). Explainable AI to understand study interest of engineering students. *Education and Information Technologies, 29*, 4657–4672.

Love, P. E. D., Fang, W., Matthews, J., Porter, S., Luo, H., & Ding, L. (2023). Explainable artificial intelligence (XAI): Precepts, models, and opportunities for research in construction. *Advanced Engineering Informatics, 57*, 102024.

Naser, M. Z. (2021). An engineer's guide to explainable artificial intelligence and interpretable machine learning: Navigating causality, forced goodness, and the false perception of inference. *Automation in Construction, 129*, 103821.

Tantithamthavorn, C. K., & Jiarpakdee, J. (2021). Explainable AI for software engineering. *2021 36th IEEE/ACM International Conference on Automated Software Engineering (ASE)*, Melbourne, Australia, 2021, 1–2.

Federated Learning

Banabilah, S., Aloqaily, M., Alsayed, E., Malik, N., & Jararweh, Y. (2022). Federated learning review: Fundamentals, enabling technologies, and future applications. *Information Processing & Management, 59*, 103061.

Gupta, R., & Gupta, J. (2023). Federated learning using game strategies: State-of-the-art and future trends. *Computer Networks, 225*, 109650.

Maroua, D. (2024). A state-of-the-art on federated learning for vehicular communications. *Vehicular Communications, 45*, 100709.

Wen, J., Zhang, Z., Lan, Y., Cui, Z., Cai, J., & Zhang, W. (2022). A Survey on federated learning: Challenges and applications. *International Journal of Machine Learning and Cybernetics, 14*, 513–535.

Generative Adversarial Network

Aggarwal, A., Mittal, M., & Battineni, G. (2021). Generative adversarial network: An overview of theory and applications. *International Journal of Information Management Data Insights, 1*, 100004.
Gonog, L., & Zhou, Y. (2019). A review: Generative adversarial networks. *2019 14th IEEE Conference on Industrial Electronics and Applications (ICIEA)*, Xi'an, China, 2019, 505–510.
Jozdani, S., Chen, D., Pouliot, D., & Johnson, B. A. (2022). A review and meta-analysis of generative adversarial networks and their applications in remote sensing. *International Journal of Applied Earth Observation and Geoinformation, 108*, 102734.
Lee, M. (2023). Recent advances in generative adversarial networks for gene expression data: A comprehensive review. *Mathematics, 11*, 3055.
Nayak, A. A., Venugopala, P. S., & Ashwini, B. A. (2024). Systematic review on generative adversarial network (GAN): Challenges and future directions. *Archives of Computational Methods in Engineering*. 31, 4739–4772.

Graph Neural Networks

Besharatifard, M., & Vafaee, F. (2024). A review on graph neural networks for predicting synergistic drug combinations. *Artificial Intelligence Review, 57*, 49.
Corso, G., Stark, H., Jegelka, S., Jaakkola, T., & Barzilay, R. (2024). Graph neural networks. *Nature Reviews Methods Primers, 4*, 17.
Khemani, B., Patil, S., Kotecha, K., & Tanwar, S. (2024). A Review of graph neural networks: Concepts, architectures, techniques, challenges, datasets, applications, and future directions. *Journal of Big Data, 11*, 18.
Sun, C., Li, C., Lin, X., Zheng, T., Meng, F., Rui, X., & Wang, Z. (2023). Attention-based graph neural networks: A survey. *Artificial Intelligence Review, 56*(Suppl 2), 2263–2310.
Zhou, J., Cui, G., Hu, S., Zhang, Z., Yang, C., Liu, Z., Wang, L., Li, C., & Sun, M. (2020). Graph neural networks: A review of methods and applications. *AI Open, 1*, 57–81.

Neural Architecture Search

Chitty-Venkata, K. T., Emani, M., Vishwanath, V., & Somani, A. K. (2023). Neural architecture search benchmarks: Insights and survey. *IEEE Access, 11*, 25217–25236.
Elsken, T., Metzen, J. H., & Hutter, F. (2019). Neural architecture search: A survey. *The. Journal of Machine Learning Research, 20*, 1997–2017.
Poyser, M., & Breckon, T. P. (2024). Neural architecture search: A contemporary literature review for computer vision applications. *Pattern Recognition, 147*, 110052.

Neural Network Pruning and Quantization

Alqahtani, A., Xie, X., & Jones, M. W. (2021). Literature review of deep network compression. *Informatics, 8*, 77.
Cai, M., Su, Y., Wang, B., & Zhang, T. (2023a). Research on compression pruning methods based on deep learning. *Journal of Physics: Conference Series, 2580*, 012060.
Liang, T., Glossner, J., Wang, L., Shi, S., & Zhang, X. (2021). Pruning and quantization for deep neural network acceleration: A survey. *Neurocomputing, 461*, 370–403.

Zhang, Y., Wang, G., Yang, T., Pang, T., He, Z., & Lv, J. (2022). Compression of deep neural networks: bridging the gap between conventional-based pruning and evolutionary approach. *Neural Computing and Applications, 34*, 16493–16514.

Neural Style Transfer

Cai, Q., Ma, M., Wang, C., & Li, H. (2023b). Image neural style transfer: A review. *Computers and Electrical Engineering, 108*, 108723.
Li, J., Wang, Q., Chen, H., An, J., & Li, S. (2020a). A review on neural style transfer. *Journal of Physics: Conference Series, 1651*, 012156.
Singh, A., Jaiswal, V., Joshi, G., Sanjeeve, A., Gite, S., & Kotecha, K. (2021). Neural style transfer: A critical review. *IEEE Access, 9*, 131583–131613.

Evolutionary Algorithms

Archimedes optimization algorithm

Dhal, K. G., Ray, S., Rai, R., & Das, A. (2023). Archimedes optimizer: Theory, analysis, improvements, and applications. *Archives of Computational Methods in Engineering, 30*, 2543–2578.
Fang, L., Yao, Y., & Liang, X. (2023). New binary archimedes optimization algorithm and its application. *Expert Systems with Applications, 230*, 120639.
Hashim, F. A., Hussain, K., Houssein, E. H., Mabrouk, M. S., & Al-Atabany, W. (2021). Archimedes optimization algorithm: A new metaheuristic algorithm for solving optimization problems. *Applied Intelligence, 51*, 1531–1551.

Artificial Algae

Turkoglu, B., Uymaz, S. A., & Kaya, E. (2022). Binary artificial algae algorithm for feature selection. *Applied Soft Computing, 120*, 108630.
Uymaz, S. A., Tezel, G., & Yel, E. (2015). Artificial algae algorithm (AAA) for non-linear global optimization. *Applied Soft Computing, 31*, 153–171.

Artificial Bee Colony Optimization

Xiao, W. S., Li, G. X., Liu, C., & Tan, L. P. (2023). A novel chaotic and neighborhood search-based artificial bee colony algorithm for solving optimization problems. *Scientific Reports, 13*, 20496.
Zhao, M., Song, X., & Xing, S. (2022). Improved artificial bee colony algorithm with adaptive parameter for numerical *Optimization Applied Artificial Intelligence, 36*. https://doi.org/10.1080/08839514.2021.2008147

Bacterial Foraging Optimization

Chen, H., Wang, L., Di, J., & Ping, S. (2020). Bacterial foraging optimization based on self-adaptive chemotaxis strategy. *Computational Intelligence and Neuroscience*, 2020, 2630104.
Guo, C., Tang, H., Niu, B., & Lee, C. B. P. (2021). A survey of bacterial foraging optimization. *Neurocomputing, 452*, 728–746.

Bat Algorithm

Gagnon, I., April, A., & Abran, A. (2020). A critical analysis of the BAT algorithm. *Engineering Reports, 2*, e12212.
Gandomi, A. H., & Yang, X. S. (2014). Chaotic bat algorithm. *Journal of Computational Science, 5*, 224–232.
Shehab, M., Abu-Hashem, M. A., Shambour, M. K. Y., Alsalibi, A. I., Alomari, O. A., Gupta, J. N. D., Alsoud, A. R., Abuhaija, B., & Abualigah, L. (2023). A comprehensive review of bat inspired algorithm: Variants, applications, and hybridization. *Archives of Computational Methods in Engineering, 30*, 765–797.

Biogeography-Based Optimization

Sang, X., Liu, X., Zhang, Z., & Wang, L. (2021). Improved biogeography-based optimization algorithm by hierarchical tissue-like P system with triggering ablation rules. *Mathematical Problems in Engineering, 2021*, 6655614.
Wei, L., Zhang, Q., & Yang, B. (2022). Improved biogeography-based optimization algorithm based on hybrid migration and dual-mode mutation strategy. *Fractal and Fractional, 6*, 597.

Black Hole Optimization

Abualigah, L., Elaziz, M. A., Sumari, P., Khasawneh, A. M., Alshinwan, M., Mirjalili, S., Shehab, M., Abuaddous, H. Y., & Gandomi, A. H. (2022a). Black hole algorithm: A comprehensive survey. *Applied Intelligence, 52*, 11892–11915.
Deeb, H., Sarangi, A., Mishra, D., & Sarangi, S. K. (2022). Improved black hole optimization algorithm for data clustering. *Journal of King Saud University—Computer and Information Sciences. Part A, 34*, 5020–5029.

Chicken Swarm Optimization Algorithm

Liang, J., Wang, L., & Ma, M. (2023). An adaptive dual-population collaborative chicken swarm optimization algorithm for high-dimensional optimization. *Biomimetics, 8*, 210.
Wang, Z., Zhang, W., Guo, Y., Han, M., Wan, B., & Liang, S. (2023a). A multiobjective chicken swarm optimization algorithm based on dual external archive with various elites. *Applied Soft Computing, 133*, 109920.
Zhang, Y., Wang, L., & Zhao, J. (2023). PECSO: An improved chicken swarm optimization algorithm with performance-enhanced strategy and its application. *Biomimetics, 8*, 355.
Zouache, D., Arby, Y. O., Nouioua, F., & Abdelaziz, F. B. (2019). Multiobjective chicken swarm optimization: A novel algorithm for solving multiobjective optimization problems. *Computers & Industrial Engineering, 129*, 377–391.

Crow Search Algorithm

Askarzadeh, A. (2016). A novel metaheuristic method for solving constrained engineering optimization problems: Crow search algorithm. *Computers & Structures, 169*, 1–12.

Hussien, A. G., Amin, M., Wang, M., Liang, G., Alsanad, A., Gumaei, A., & Chen, H. (2020). Crow search algorithm: Theory, recent advances, and applications. *IEEE Access, 8*, 173548–173565.

Cuckoo Search Algorithm

Mohamad, A. B., Zain, A. M., & Bazin, N. E. N. (2014). Cuckoo search algorithm for optimization problems—a literature review and its applications. *Applied Artificial Intelligence, 28*, 419–448.

Xiong, Y., Zou, Z., & Cheng, J. (2023). Cuckoo search algorithm based on cloud model and its application. *Scientific Reports, 13*, 10098.

Elephant Clan Optimization

Jafari, M., Salajegheh, E., & Salajegheh, J. (2021). Elephant clan optimization: a nature-inspired metaheuristic algorithm for the optimal design of structures. *Applied Soft Computing*, 113, Part A, 107892.

Elephant Herding Optimization

Li, J., Lei, H., Alavi, A. H., & Wang, G. G. (2020b). Elephant herding optimization: Variants, hybrids, and applications. *Mathematics, 8*, 1415.

Equilibrium Optimizer

Elmanakhly, D. A., Saleh, M. M., & Rashed, E. A. (2021). An improved equilibrium optimizer algorithm for features selection: Methods and analysis. *IEEE Access, 9*, 120309–120327.

Faramarzi, A., Heidarinejad, M., Stephens, B., & Mirjalili, S. (2020). Equilibrium optimizer: A novel optimization algorithm. *Knowledge-Based Systems, 191*, 105190.

Yang, L., Xu, Z., Liu, Y., & Tian, G. (2022a). An improved equilibrium optimizer with a decreasing equilibrium pool. *Symmetry, 14*, 1227.

Fish Swarm

Pourpanah, F., Wang, R., Lim, C. P., Wang, X. Z., & Yazdani, D. (2023). A review of artificial fish swarm algorithms: Recent advances and applications. *Artificial Intelligence Review, 56*, 1867–1903.

Tan, W. H., & Mohamad-Saleh, J. (2020). Normative fish swarm algorithm (NFSA) for optimization. *Soft Computing, 24*, 2083–2099.

Gravitational Search Algorithm

Mittal, H., Tripathi, A., Pandey, A. C., & Pal, R. (2021). Gravitational search algorithm: A comprehensive analysis of recent variants. *Multimedia Tools and Applications, 80*, 7581–7608.

Rashedi, E., Nezamabadi-pour, H., & Saryazdi, S. (2009). GSA: A gravitational search algorithm. *Information Sciences, 179*, 2232–2248.

Yang, Z., Cai, Y., & Li, G. (2022b). Improved gravitational search algorithm based on adaptive strategies. *Entropy, 24*, 1826.

Grey Wolf Optimization Algorithm

Mirjalili, S., Mirjalili, S. M., & Lewis, A. (2014). Grey wolf optimizer. *Advances in Engineering Software, 69*, 46–61.

Pan, C., Si, Z., Du, X., & Lv, Y. (2021). A four-step decision-making grey wolf optimization algorithm. *Soft Computing, 25*, 14375–14391.

Wang, J. S., & Li, S. X. (2019). An improved grey wolf optimizer based on differential evolution and elimination mechanism. *Scientific Reports, 9*, 7181.

Harmony Search

Kim, J. H. (2016). Harmony search algorithm: A unique music-inspired algorithm. *Procedia Engineering, 154*, 1401–1405.

Wang, J., Ouyang, H., Zhang, C., Li, S., & Xiang, J. (2023b). A novel intelligent global harmony search algorithm based on improved search stability strategy. *Scientific Reports, 13*, 7705.

Henry Gas Solubility Optimization

Hashim, F. A., Houssein, E. H., Mabrouk, M. S., Al-Atabany, W., & Mirjalili, S. (2019). Henry gas solubility optimization: A novel physics-based algorithm. *Future Generation Computer Systems, 101*, 646–667.

Li, S., Wang, J. S., Xie, W., & Li, X. L. (2022b). An improved Henry gas solubility optimization algorithm based on lévy flight and brown motion. *Applied Intelligence, 52*, 12584–12608.

Mohammadi, D., Abd Elaziz, M., Moghdani, R., Demir, E., & Mirjalili, S. (2022). Quantum henry gas solubility optimization algorithm for global optimization. *Engineering with Computers, 38*(Suppl 3), 2329–2348.

Honey Badger Algorithm

Hashim, F. A., Houssein, E. H., Hussain, K., Mabrouk, M. S., & Al-Atabany, W. (2022). Honey badger algorithm: New metaheuristic algorithm for solving optimization problems. *Mathematics and Computers in Simulation, 192*, 84–110.

Imperialist Competitive Algorithm

Abdollahi, M., Isazadeh, A., & Abdollahi. (2013). Imperialist competitive algorithm for solving systems of non-linear equations. *Computers & Mathematics with Applications, 65*, 1894–1908.

Bernal, E., Castillo, O., Soria, J., & Valdez, F. (2017). Imperialist competitive algorithm with dynamic parameter adaptation using fuzzy logic applied to the optimization of mathematical functions. *Algorithms, 10*, 18.

Hosseini, S., & Al Khaled, A. (2014). A survey on the imperialist competitive algorithm metaheuristic: Implementation in engineering domain and directions for future research. *Applied Soft Computing, 24*, 1078–1094.

Indicator-Based Evolutionary Algorithm

García, J. L. L., Monroy, R., Hernández, V. A. S., & Coello, C. A. C. (2021). COARSE-EMOA: An indicator-based evolutionary algorithm for solving equality constrained multiobjective optimization problems. *Swarm and Evolutionary Computation, 67*, 100983.

Li, F., Yang, Y., Shang, Z., Li, S., & Ouyang, H. (2023b). Kriging-assisted indicator-based evolutionary algorithm for expensive multiobjective optimization. *Applied Soft Computing, 147*, 110736.

Yuan, J., Liu, H. L., Ong, Y. S., & He, Z. (2022). Indicator-based evolutionary algorithm for solving constrained multiobjective optimization problems. *IEEE Transactions on Evolutionary Computation, 26*, 379–391.

Jellyfish Search Optimization

Chou, J. S., & Molla, A. (2022). Recent advances in use of bio-inspired jellyfish search algorithm for solving optimization problems. *Scientific Reports, 12*, 19157.

Chou, J. S., & Truong, D. N. (2021). A novel metaheuristic optimizer inspired by behavior of jellyfish in ocean. *Applied Mathematics and Computation, 389*, 125535.

Manita, G., & Zermani, A. (2021). A modified jellyfish search optimizer with orthogonal learning strategy. *Procedia Computer Science, 192*, 697–708.

Krill Herd Algorithm

Bolaji, A. L., Al-Betar, M. A., Awadallah, M. A., Khader, A. T., & Abualigah, L. M. (2016). A comprehensive review: Krill Herd Algorithm (KH) and its applications. *Applied Soft Computing, 49*, 437–446.

Gandomi, A. H., & Alavi, A. H. (2012). Krill Herd: A new bio-inspired optimization algorithm. *Communications in Non-linear Science and Numerical. Simulation, 17*, 4831–4845.

Wang, G. G., Gandomi, A. H., Alavi, A. H., & Gong, D. (2019). A comprehensive review of Krill Herd algorithm: Variants. *Hybrids and Applications, Artificial Intelligence Review, 51*, 119–148.

Learner performance-based behavior algorithm

Rahman, C. M., & Rashid, T. A. (2021). A new evolutionary algorithm: learner performance-based behavior algorithm. *Egyptian Informatics Journal, 22*, 213–223.

Meta-Heuristic Algorithms

Abdel-Basset, M., Abdel-Fatah, L., & Sangaiah, A. K. (2018). Metaheuristic algorithms: A comprehensive review. In A. K. Sangaiah, M. Sheng, & Z. Zhang (Eds.), *Intelligent Data-Centric Systems, Computational Intelligence for Multimedia Big Data on the Cloud with Engineering Applications*. Academic Press, 185–231.

Alorf, A. (2023). A survey of recently developed metaheuristics and their comparative analysis. *Engineering Applications of Artificial Intelligence*, 117 (Part A), 105622.

Khanduja, N., & Bhushan, B. (2020). Recent advances and application of metaheuristic algorithms: A survey (2014–2020). *Metaheuristic and Evolutionary Computation: Algorithms and Applications, Part of the Studies in Computational Intelligence Book Series, 916*, 207–228.

Rajalakshmi, S., & Kanmani, S. (2022). A comprehensive review on recent intelligent metaheuristic algorithms. *International Journal of Swarm Intelligence, 7*, 182–205.

Rajwar, K., Deep, K., & Das, S. (2023). An exhaustive review of the metaheuristic algorithms for search and optimization: Taxonomy. *Applications, and Open Challenges, Artificial Intelligence Review, 56*, 13187–13257.

Reddy, M. J., & Kumar, D. N. (2020). Evolutionary algorithms, swarm intelligence methods, and their applications in water resources engineering: A state-of-the-art review. *H2Open Journal*, 3, 135–188 [open access paper under CC BY- 4.0 License].

Velasco, L., Guerrero, H., & Hospitaler, A. A. (2024). A literature review and critical analysis of metaheuristics recently developed. *Archives of Computational Methods in Engineering, 31*, 125–146.

Wong, W., & Ming, C. I. (2019). A review on metaheuristic algorithms: Recent trends, benchmarking and applications. *7th International Conference on Smart Computing & Communications (ICSCC)*, Sarawak, Malaysia, 2019, 1–5.

Moth Flame Optimization

Mirjalili, S. (2015). Moth-flame optimization algorithm: A novel nature-inspired heuristic paradigm. *Knowledge-Based Systems, 89*, 228–249.

Nadimi-Shahraki, M. H., Zamani, H., Fatahi, A., & Mirjalili, S. (2023a). MFO-SFR: An enhanced moth-flame optimization algorithm using an effective stagnation finding and replacing strategy. *Mathematics, 11*, 862.

Sahoo, S. K., Saha, A. K., Ezugwu, A. E., Agushaka, J. O., Abuhaija, B., Alsoud, A. R., & Abualigah, L. (2023). Moth flame optimization: Theory, modifications, hybridizations, and applications. *Archives of Computational Methods in Engineering, 30*, 391–426.

Shehab, M., Abualigah, L., Al Hamad, H., Alabool, H., Alshinwan, M., & Khasawneh, A. M. (2020). Moth-flame optimization algorithm: Variants and applications. *Neural Computing and Applications, 32*, 9859–9884.

Mountain Gazelle Optimizer

Abdollahzadeh, B., Gharehchopogh, F. S., Khodadadi, N., & Mirjalili, S. (2022). Mountain gazelle optimizer: A new nature-inspired metaheuristic algorithm for global optimization problems. *Advances in Engineering Software, 174*, 103282.

Mountaineering Team-Based Optimization

Faridmehr, I., Nehdi, M. L., Davoudkhani, I. F., & Poolad, A. (2023). Mountaineering team-based optimization: A novel human-based metaheuristic algorithm. *Mathematics, 11*, 1273.

Multiverse Optimization Algorithm

Benmessahel, I., Xie, K., & Chellal, M. (2020). A new competitive multiverse optimization technique for solving single-objective and multiobjective problems. *Engineering Reports, 2*, e12124.

Mirjalili, S., Mirjalili, S. M., & Hatamlou, A. (2016). Multi-verse optimizer: A nature-inspired algorithm for global optimization. *Neural Computing and Applications, 27*, 495–513.

Red Fox Optimization

Połap, D., & Woźniak, M. (2021). Red Fox optimization algorithm. *Expert Systems with Applications, 166*, 114107.

Reptile Search Algorithm

Abualigah, L., Elaziz, M. A., Sumari, P., Geem, Z. W., & Gandomi, A. H. (2022b). Reptile search algorithm (RSA): A nature-inspired meta-heuristic optimizer. *Expert Systems with Applications, 191*, 116158.

Khan, M. K., Zafar, M. H., Rashid, S., Mansoor, M., Moosavi, S. K. R., & Sanfilippo, F. (2023). Improved reptile search optimization algorithm: Application on regression and classification problems. *Applied Sciences, 13*, 945.

Sasmal, B., Hussien, A. G., Das, A., Dhal, K. G., & Saha, R. (2024). Reptile search algorithm: Theory, variants, applications, and performance evaluation. *Archives of Computational Methods in Engineering, 31*, 521–549.

Salp Swarm Algorithm

Duan, Q., Wang, L., Kang, H., Shen, Y., Sun, X., & Chen, Q. (2021). Improved salp swarm algorithm with simulated annealing for solving engineering optimization problems. *Symmetry, 13*, 1092.

Hegazy, Ah. E., Makhlouf, M. A., & El-Tawel, Gh. S. (2020). Improved salp swarm algorithm for feature selection. *Journal of King Saud University—Computer and Information Sciences, 32*, 335–344.

Mirjalili, S., Gandomi, A. H., Mirjalili, S. Z., Saremi, S., Faris, H., & Mirjalili, S. M. (2017). Salp swarm algorithm: A bio-inspired optimizer for engineering design problems. *Advances in Engineering Software, 114*, 163–191.

Sand Cat swarm optimization

Kiani, F., Nematzadeh, S., Anka, F. A., & Findikli, M. A. (2023). Chaotic sand cat swarm optimization. *Mathematics, 11*, 2340.

Seyyedabbasi, A., & Kiani, F. (2023). Sand cat swarm optimization: A nature-inspired algorithm to solve global optimization problems. *Engineering with Computers, 39*, 2627–2651.
Wu, D., Rao, H., Wen, C., Jia, H., Liu, Q., & Abualigah, L. (2022). Modified sand cat swarm optimization algorithm for solving constrained engineering optimization problems. *Mathematics, 10*, 4350.

Shuffled frog leaping algorithm

Maaroof, B. B., Rashid, T. A., Abdulla, J. M., Hassan, B. A., Alsadoon, A., Mohammadi, M., Khishe, M., & Mirjalili, S. (2022). Current studies and applications of shuffled frog leaping algorithm: A review. *Archives of Computational Methods in Engineering, 29*, 3459–3474.
Zhao, Z., Wang, M., Liu, Y., Chen, Y., He, K., & Liu, Z. (2024). A modified shuffled frog leaping algorithm with inertia weight. *Scientific Reports, 14*, 4146.

Social Spider Optimization

Feng, S., Hu, Y., Chen, Y., & Huang, M. (2022). An improved spider optimization algorithm coordinated by pheromones. *Scientific Reports, 12*, 5962.
Yu, J. J. Q., & Vok, L. (2015). A social spider algorithm for global optimization. *Applied Soft Computing, 30*, 614–627
Zhao, R., Luo, Q., & Zhou, Y. (2017). Elite opposition-based social spider optimization algorithm for global function optimization. *Algorithms, 10*, 9.

Spider Monkey Optimization Algorithm

Agrawal, A., Garg, D., Popli, D., Banerjee, A., Raj, A., & Dikshit, I. (2023). A review of spider monkey optimization: Modification and its biomedical application. *International Journal on Interactive Design and Manufacturing*. https://doi.org/10.1007/s12008-023-01671-4

Swarm Intelligence

Brezočnik, L., Fister, I., & Podgorelec, V. (2018). Swarm intelligence algorithms for feature selection: A review. *Applied Sciences, 8*, 1521.
Figueiredo, E., Macedo, M., Siqueira, H. V., Santana, C. J., Gokhale, A., & Bastos-Filho, C. J. A. (2019). Swarm intelligence for clustering—A systematic review with new perspectives on data mining. *Engineering Applications of Artificial Intelligence, 82*, 313–329.
Tang, J., Liu, G., & Pan, Q. (2021). A review on representative swarm intelligence algorithms for solving optimization problems: applications and trends. *IEEE/CAA Journal of Automatica Sinica, 8*, 1627–1643.

Symbiotic Organisms Search

Cheng, M. Y., & Prayogo, D. (2014). Symbiotic organisms search: A new metaheuristic optimization algorithm. *Computers & Structures, 139*, 98–112.

Ezugwu, A. E., & Prayogo, D. (2019). Symbiotic organisms search algorithm: Theory, recent advances and applications. *Expert Systems with Applications, 119*, 184–209.

Gharehchopogh, F. S., Shayanfar, H., & Gholizadeh, H. (2020). A comprehensive survey on symbiotic organisms search algorithms. *Artificial Intelligence Review, 53*, 2265–2312.

Teaching-Learning-Based Optimization

Gómez Díaz, K. Y., De León Aldaco, S. E., Aguayo Alquicira, J., Ponce-Silva, M., & Olivares Peregrino, V. H. (2022). Teaching–learning-based optimization algorithm applied in electronic engineering: A survey. *Electronics, 11*, 3451.

Xu, Y., Peng, Y., Su, X., Yang, Z., Ding, C., & Yang, X. (2022). Improving teaching–learning-based-optimization algorithm by a distance-fitness learning strategy. *Knowledge-Based Systems, 257*, 108271.

Teamwork Optimization Algorithm

Dehghani, M., & Trojovský, P. (2021). Teamwork optimization algorithm: A new optimization approach for function minimization/maximization. *Sensors, 21*, 4567.

Walrus Optimization Algorithm

Han, M., Du, Z., Yuen, K. F., Zhu, H., Li, Y., & Yuan, Q. (2024). Walrus optimizer: A novel nature-inspired metaheuristic algorithm. *Expert Systems with Applications, 239*, 122413.

Whale Optimization Algorithm

Mirjalili, S., & Lewis, A. (2016). The whale optimization algorithm. *Advances in Engineering Software, 95*, 51–67.

Nadimi-Shahraki, M. H., Zamani, H., Varzaneh, Z. A., & Mirjalili, S. (2023b). A systematic review of the whale optimization algorithm: Theoretical foundation, improvements, and hybridizations. *Archives of Computational Methods in Engineering, 30*, 4113–4159.

Rana, N., Latiff, M. S. A., Abdulhamid, S. M., & Chiroma, H. (2020). Whale optimization algorithm: A systematic review of contemporary applications. *Modifications and Developments, Neural Computing and Applications, 32*, 16245–16277.

Wild Horse Optimizer

Naruei, I., & Keynia, F. (2022). Wild horse optimizer: A new meta-heuristic algorithm for solving engineering optimization problems. *Engineering with Computers, 38*(Suppl 4), 3025–3056.

Zheng, R., Hussien, A. G., Jia, H.-M., Abualigah, L., Wang, S., & Wu, D. (2022). An improved wild horse optimizer for solving optimization problems. *Mathematics, 10*, 1311.

Benchmark (or test) Functions used in Optimization algorithms

Hellwig, M., & Beyer, H. G. (2019). Benchmarking evolutionary algorithms for single objective real-valued constrained optimization—a critical review. *SWarm and Evolutionary Computation, 44*, 927–944.

Molga, M., & Smutnicki, C. (2005). Test Functions for Optimization Needs. https://robertmarks.org/Classes/ENGR5358/Papers/functions.pdf, Accessed October 1, 2024

Piotrowski, A. P., Napiorkowski, J. J., & Piotrowska, A. E. (2023). Choice of benchmark optimization problems does matter. *Swarm and Evolutionary Computation, 83*, 101378.

Surjanovic, S., & Bingham, D. (2013). *Virtual Library of Simulation Experiments: Test Functions and Datasets*. Available at http://www.sfu.ca/~ssurjano, Accessed October 29, 2023

Volz, V., Irawan, D., van der Blom, K., & Naujoks, B. (2023). Benchmarking. In D. Brockhoff, M. Emmerich, B. Naujoks, & R. Purshouse (Eds.), *Many-Criteria Optimization and Decision Analysis, Natural Computing Series*, Springer, Cham.

Zitzler, E., Deb, K., & Thiele, L. (2000). Comparison of multiobjective evolutionary algorithms: Empirical results. *Evolutionary Computation, 8*, 125–148.

Artificial Intelligence Tools

Agarwal, M., Sharma, P., & Goswami, A. (2023). Analysing the applicability of ChatGPT, bard, and bing to generate reasoning-based multiple-choice questions in medical physiology. *Cureus, 15*, e40977.

Cheong, R. C. T., Unadkat, S., Mcneillis, V., Williamson, A., Joseph, J., Randhawa, P., Andrews, P., & Paleri, V. (2023). Artificial intelligence chatbots as sources of patient education material for obstructive sleep apnoea: ChatGPT versus google bard. *European Archives of Oto-Rhino-Laryngology, 281*, 985–993.

Dempere, J., Modugu, K., Hesham, A., & Ramasamy, L. K. (2023). The impact of ChatGPT on higher education. *Frontiers in Education*, 8–2023.

Foroumandi, E., Moradkhani, H., Sanchez-Vila, X., Singha, K., Castelletti, A., & Destouni, G. (2023). ChatGPT in hydrology and earth sciences: Opportunities, prospects, and concerns. *Water Resources Research*, 59, e2023WR036288.

Huang, J., & Tan, M. (2023). The role of ChatGPT in scientific communication: writing better scientific review articles. *American Journal of Cancer Research, 13*, 1148–1154.

Lim, B., Seth, I., Bulloch, G., Xie, Y., Hunter-Smith, D. J., & Rozen, W. M. (2023). Evaluating the efficacy of major language models in providing guidance for hand trauma nerve laceration patients: A case study on google's AI BARD, Bing AI, and ChatGPT. *Plastic and Aesthetic Research, 10*, 43.

McGowan, A., Gui, Y., Dobbs, M., Shuster, S., Cotter, M., Selloni, A., Goodman, M., Srivastava, A., Cecchi, G. A., & Corcoran, C. M. (2023). ChatGPT and bard exhibit spontaneous citation fabrication during psychiatry literature search. *Psychiatry Research, 326*, 115334.

Nah, F. F., Zheng, R., Cai, J., Siau, K., & Chen, L. (2023). Generative AI and ChatGPT: Applications, challenges, and ai-human collaboration. *Journal of Information Technology Case and Application Research, 25*, 277–304.

Nikolic, S., Daniel, S., Haque, R., Belkina, M., Hassan, G. M., Grundy, S., Lyden, S., Neal, P., & Sandison, C. (2023). ChatGPT versus engineering education assessment: A multidisciplinary and multi-institutional benchmarking and analysis of this generative artificial intelligence tool to investigate assessment integrity. *European Journal of Engineering Education, 48*, 559–614.

Ray, P. P. (2023). ChatGPT: A comprehensive review on background, applications, key challenges, bias, ethics, limitations and future scope. *Internet of Things and Cyber-Physical Systems, 3*, 121–154.

Ray, S. S., Peddinti, P. R. T., Verma, R. K., Puppala, H., Kim, B., Singh, A., & Kwon, Y. N. (2024). Leveraging ChatGPT and bard: What does it convey for water treatment/desalination and harvesting sectors? *Desalination, 570*, 117085.

Roumeliotis, K. I., & Tselikas, N. D. (2023). ChatGPT and open-AI models: A preliminary review. *Future Internet, 15*, 192.

IoT, DSS, Big Data

Alamanos, A., Rolston, A., & Papaioannou, G. (2021). Development of a decision support system for sustainable environmental management and stakeholder engagement. *Hydrology, 8*, 40.

Alshami, A., Elsayed, M., Ali, E., Eltoukhy, A. E. E., & Zayed, T. (2023). Harnessing the power of ChatGPT for automating systematic review process: Methodology, case study, limitations, and future directions. *Systems, 11*, 351.

Gheibi, M., Taghavian, H., Moezzi, R., Waclawek, S., Cyrus, J., Dawiec-Lisniewska, A., Koci, J., & Khaleghiabbasabadi, M. (2023). Design of a decision support system to operate a NO_2 gas sensor using machine learning, sensitive analysis and conceptual control process modelling. *Chemosensors, 11*, 126.

Jena, S. (2023). Architecture of Internet of Things (IoT), https://www.geeksforgeeks.org/architecture-of-internet-of-things-iot/. Accessed 26 July 2023

Kapliński, O., Košeleva, N., & Ropaitė, G. (2016). Big data in civil engineering: A state-of-the-art survey. *Engineering Structures and Technologies, 8*, 165–175.

Thayyib, P. V., Mamilla, R., Khan, M., Fatima, H., Asim, M., Anwar, I., Shamsudheen, M. K., & Khan, M. A. (2023). State-of-the-art of artificial intelligence and big data analytics reviews in five different domains: A bibliometric summary. *Sustainability, 15*, 4026.

Taguchi Design of Experiments

Chen, W. H., Wu, D. R., Chang, M. H., Rajendran, S., Ong, H. C., & Lin, K. Y. A. (2024). Modeling of hydrogen separation through PD membrane with vacuum pressure using taguchi and machine learning methods. *International Journal of Hydrogen Energy*.https://doi.org/10.1016/j.ijhydene.2024.08.204

Cimbala, J. M. (2014). Taguchi Orthogonal Arrays (Accessed at https://www.me.psu.edu/cimbala/me345/Lectures/Taguchi_orthogonal_arrays.pdf on 2.9.2024)

Fraley, S., Zalewski, J., Oom, M., & Terrien, B. (2024). Design of Experiments via Taguchi Methods—Orthogonal Arrays (Article is available at https://eng.libretexts.org/Bookshelves/Industrial_and_Systems_Engineering/Chemical_Process_Dynamics_and_Controls_(Woolf)/14%3A_Design_of_Experiments/14.01%3A_Design_of_Experiments_via_Taguchi_Methods_-_Orthogonal_Arrays). Accessed September 1, 2024

Frey, D. (1998). Constructing Orthogonal Arrays (Accessed at https://ocw.mit.edu/courses/16-881-robust-system-design-summer-1998/resources/l8_orth_arrays/ on 04.09.2024). Also referred https://ocw.mit.edu/courses/16-881-robust-system-design-summer-1998/pages/lecture-notes/ on 04.09.2024)

Ginting, E., & Tambunan, M. M. (2018). Selection of optimal factor level from process parameters in palm oil industry. *The 2nd Annual Applied Science and Engineering Conference (AASEC 2017), IOP Conf. Series: Materials Science and Engineering* 288, 012056.

Kacker, R. N., Lagergren, E. S., & Filliben, J. J. (1991). Taguchi's orthogonal arrays are classical designs of experiments. *Journal of Research of the National Institute of Standards and Technology, 96*, 577–591.

Li, H., Zeng, Q., Zhuang, Y., Wang, Y., Ye, Z., & Cui, (2024). Multi-step optimization of hybrid cooling array via integrating taguchi method and long short-term memory neural network. *Applied Thermal Engineering, 254*,123814.

Nejati, E., Ghaedy-Heidary, E., Ghasemi, A., & Torabi, S. A. (2024). A machine learning-based simulation metamodeling method for dynamic scheduling in smart manufacturing systems. *Computers & Industrial Engineering, 196*, 110507.

Data Augmentation

Chen, H., Birkelund, Y., & Zhang, Q. (2021). Data-augmented sequential deep learning for wind power forecasting. *Energy Conversion and Management, 248*, 114790.

Garcea, F., Serra, A., Lamberti, F., & Morra, L. (2023). Data augmentation for medical imaging: A systematic literature review. *Computers in Biology and Medicine, 152*, 106391.

Goceri, E. (2023). Medical image data augmentation: Techniques, comparisons and interpretations. *Artificial Intelligence Review, 56*, 12561–12605.

Mahlathi, C. D., Wilms, J., & Brink, I. (2022). Investigation of scarce input data augmentation for modelling nitrogenous compounds in South African rivers. *Water Practice and Technology, 17*, 2499–2515.

Ranjbaran, G., Reforgiato Recupero, D., & Lombardo, G. Consoli, S. (2023) leveraging augmentation techniques for tasks with unbalancedness within the financial domain: A two-level ensemble approach. *EPJ Data Science, 12*, 24.

Cross-Validation

Kaliappan, J., Bagepalli, A. R., Almal, S., Mishra, R., Hu, Y. C., & Srinivasan, K. (2023). Impact of cross-validation on machine learning models for early detection of intrauterine fetal demise. *Diagnostics (Basel), 13*, 1692.

Kee, E., Chong, J. J., Choong, Z. J., & Lau, M. (2023). A comparative analysis of cross-validation techniques for a smart and lean pick-and-place solution with deep learning. *Electronics, 12*, 2371.

Madhuri, R., Sistla, S., & Raju, K. S. (2021). Application of machine learning algorithms for flood susceptibility assessment and risk management. *Journal of Water and Climate Change, 12*, 2608–2623.

Tougui, I., Jilbab, A., & Mhamdi, J. E. (2021). Impact of the choice of cross-validation techniques on the results of machine learning-based diagnostic applications. *Healthcare Informatics Research, 27*, 189–199.

Suggested Further Reading

Abdelmoneim, A. A., Khadra, R., Derardja, B., & Dragonetti, G. (2023). Internet of things (IoT) for soil moisture tensiometer automation. *Micromachines, 14*, 263.

Al-Ghamdi, R., & Sharma, S. K. (2022). IoT-based smart water management systems for residential buildings in Saudi Arabia. *Processes, 10*, 2462.

Al-Metwally, S. A. H., Hassan, M. K., & Mourad, M. H. (2020). Real Time internet of things (IoT) based water quality management system. *Procedia CIRP, 91*, 478–485.

AquaBit. (2018). Using Blockchain Technology to Adaptively Manage Water Resources, http://aquabit.io/. Accessed June 26, 2023.

Entezami, A., Sarmadi, H., Behkamal, B., & Mariani, S. (2020). Big data analytics and structural health monitoring: A statistical pattern recognition-based approach. *Sensors, 20*, 2328.

Hou, Y., Li, Q., Zhang, C., Lu, G., Ye, Z., Chen, Y., Wang, L., & Cao, D. (2021). The state-of-the-art review on applications of intrusive sensing, image processing techniques, and machine learning methods in pavement monitoring and analysis. *Engineering, 7*, 845–856.

Hyperledger. (2018). Sawtooth: An Introduction https://www.hyperledger.org/learn/white-papers. Accessed June 26, 2023

Khurshid, K., Danish, A., Salim, M. U., Bayram, M., Ozbakkaloglu, T., & Mosaberpanah, M. A. (2023). An In-depth survey demystifying the internet of things (IoT) in the construction industry: unfolding new dimensions. *Sustainability, 15*, 1275.

Minhas, M. R., & Potdar, V. (2020). Decision support systems in construction: A bibliometric analysis. *Buildings, 10*, 108.

Muazu, A. A., Hashim, A. S., & Sarlan, A. (2022). Review of nature inspired metaheuristic algorithm selection for combinatorial T-way testing. *IEEE Access, 10*, 27404–27431.

Shaikh, P. W., El-Abd, M., Khanafer, M., & Gao, K. (2022). A review on swarm intelligence and evolutionary algorithms for solving the traffic signal control problem. *IEEE Transactions on Intelligent Transportation Systems, 23*, 48–63.

Shashank. (2023). Top 55 Blockchain Interview Questions You Must Prepare In 2023 (https://www.edureka.co/blog/interview-questions/Blockchain-interview-questions/. Accessed 29 June, 2023

Singh, M., & Ahmed, S. (2021). IoT based smart water management systems: A systematic review. *Materials Today: Proceedings, 46*, 5211–5218.

Slowik, A., & Kwasnicka, H. (2020). Evolutionary algorithms and their applications to engineering problems. *Neural Computing and Applications, 32*, 12363–12379.

Chapter 7
Case Studies

7.1 Introduction

The present chapter provides brief information about representative case studies related to Machine Learning, Fuzzy Inference Systems, Neuro-Fuzzy Inference Systems, Fuzzy Cognitive Mapping, Fuzzy Cluster Analysis, optimization, fuzzy extensions, and many more, demonstrating the potential of these techniques in various domains. In addition, areas of further research work are also part of this chapter.

The chosen research papers are organized into Civil, Chemical, Mechanical, Electronics and Computer Science Engineering, and Management (Fig. 7.1) and presented in Tables 7.1, 7.2, 7.3, 7.4, and 7.5, respectively.

The first column in each table provides information about the authors. The second column briefly discusses the application, techniques, and performance measures. The third column focuses mainly on case studies and data sources. The fourth column presents remarks/inferences for a comprehensive view of the research work. Wherever the number of inputs or outputs is more in the studied research paper, information about a smaller number of inputs and outputs is only presented. It is represented as inputs (representative) or outputs (representative).

The full paper can be accessed if the institution has a journal subscription or the journal is in the open-access category. Note that a very brief description is provided in this chapter based on the authors' understanding of the work, which may not be logical sometimes. Readers are strongly advised to study the paper before applying it to their research work or further learning. In addition, several state-of-art review papers on the topics are presented in the appendix. Notations were expanded whenever they appeared for the first time in the text. Later, only notations were given to minimize repetition. In addition, notations were presented at the start of the book for a comprehensive understanding of the reader.

K. Srinivasa Raju and D. Nagesh Kumar, *Artificial Intelligence and Machine Learning Techniques in Engineering and Management*,
https://doi.org/10.1007/978-981-96-2621-2_7

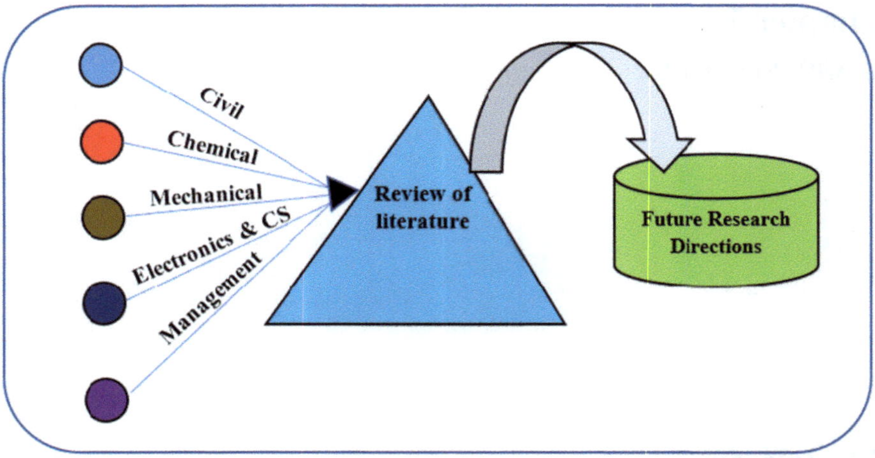

Fig. 7.1 Organization of the chapter

Table 7.1 Case studies related to Civil Engineering

Author(s) (1)	Application, techniques, and performance measures employed (2)	Case study, data source (3)	Remarks/inferences, (4)
Sujith & Jeon (2019)	Failure mode identification of bridge columns K-Nearest Neighbourhood (KNN), Naïve Bayes (NB), Quadratic Discriminant Analysis, Random Forest (RF), Artificial Neural Networks (ANN), Decision Tree (DT) Confusion Matrix	311 specimens based on laboratory studies	Inputs: Axial load, aspect, longitudinal reinforcement as well as transverse reinforcement ratios Output: Failure mode Tenfold cross-validation Categorization based on ANN is 91% accurate while assessing failure mode
Feng et al. (2021)	Shear strength prediction of concrete walls XGBoost, SHapley Additive exPlanations (SHAP), ANN, DT, Gradient Boosted Decision Tree (GBDT), RF Mean Absolute Percentage Error (MAPE), R^2, Mean Absolute Error (MAE), Root Mean Square Error (RMSE)	434 datasets of concrete walls from the shear strength database	Inputs (representative): Height, length, web thickness, flange height, flange thickness, compressive strength of Concrete, Applied axial load Output: Shear strength Tenfold cross-validation XGBoost is the best among the employed
Zhu & Wang (2021)	Deterioration prediction of bridges Recurrent Neural Network (RNN)-Convolutional Neural Network (CNN)-improved ReliefF algorithm Mean Square Error (MSE), True Accuracy Index, Conservative Accuracy Index, and Time for One Epoch	Database of 23,104 bridges selected in Texas from National Bridge Inventory	Feature selection using improved ReliefF algorithm Inputs (representative): Geographical position of bridges, average daily traffic, length of maximum span and structure, deck structure type and its width, deck protection, deck area, age, age from reconstruction, type of wearing surface and membrane, present and future average daily truck traffic, toll, lanes on structure, bridge roadway width Output: Deterioration prediction of the bridges RNN-CNN accurately predicted the condition ratings of bridges

(continued)

Table 7.1 (continued)

Author(s) (1)	Application, techniques, and performance measures employed (2)	Case study, data source (3)	Remarks/inferences, (4)
Mangalathu et al. (2022)	Seismic performance assessment Support Vector Regression (SVR), RF, Ridge Regression, KNN, DT, XGBoost, AdaBoost, SHAP R^2, RMSE	Bridge damage assessment database, California	Inputs (representative): Aspect ratio of wall, Concrete compressive strength Output: Shear strength of wall XGBoost is the best among the employed
Kourehpaz & Hutt (2022)	Enhanced regional seismic risk assessments Logistic Regression (LR), KNN, DT, RF, AdaBoost, Gradient Boosting (GB) Accuracy, Precision, F1-score, Recall	18,000 non-linear structural analysis results	Inputs: Fundamental period, wall moment of inertia, wall longitudinal reinforcement, average spectral acceleration, significant duration, displacement spectrum intensity, spectrum intensity Output: Classification of expected damage states All ML techniques performed well; GB is the most efficient, with an F1-score of 87%
Chen et al. (2022b)	Performance prediction of structures NGBoost R^2, RMSE, Confusion Matrix, Accuracy	480 highway bridges	Inputs: Concrete strength, horizontal and vertical web reinforcement strengths, and corresponding ratios; longitudinal reinforcement strength and ratio; wall height, length, flange width, thickness, axial load, web thickness Output: Shear strength Tenfold cross-validation NGBoost performed satisfactorily
Ozelim et al. (2022)	Structural health monitoring of dams Deep Neural Autoencoders—Cumulative Sum Algorithm-Fuzzy Logic Classification code, Principal Component Analysis (PCA)	Experimental setup in GeoFluxo Laboratory, University of Brasilia, Darcy Ribeiro campus	Input: Acoustic signals Output: Internal erosion in rockfill dams and body of earth The proposed approach is helpful

(continued)

Table 7.1 (continued)

Author(s) (1)	Application, techniques, and performance measures employed (2)	Case study, data source (3)	Remarks/inferences, (4)
Prieto et al. (2017)	Life of architectural buildings Mamdani Fuzzy Inference System (FIS)	Vulnerability as well as risk evaluation of five buildings situated in southern Spain and nine theoretical case studies	Inputs: Roof design, geological location, constructive system, built context, preservation, load state modification, fire, facilities, ventilation, overloads, occupancy, inner environment, heritage, population growth, furniture value, temperature, rainfall Output: Building serviceability More than 97 fuzzy rules were developed
Diker & Erkan (2022)	Design of window Mamdani FIS	8-classroom primary school design, 30 architects	Inputs: Ratio of area of window to floor (and wall), window direction Output: Window design efficiency of classrooms 27 fuzzy rules were developed
Madani et al. (2020)	Compressive strength prediction of nano-silica-incorporated cement mixtures ANN, regression-based techniques, Adaptive Network-based Fuzzy Inference System (ANFIS) R^2, RMSE, MAE, MAPE	Experimental setup, 32 mortar mixes	Inputs: Cement, age, aggregate, water, content of nano-silica and its mean size, superplasticizer Output: Compressive strength ANN and ANFIS outperformed the regression-based techniques
Khanzadi et al. (2017)	Change orders in construction projects Fuzzy Cognitive Mapping (FCM)	Underpass construction project, city of Mehriz, Yazd, Iran	Considered 17 causes Effect: Change orders in construction projects The MF corresponds to five linguistic variables Activation function: Sigmoid
Vijayalakshmi et al. (2017)	Categorization of cracks in columns of reinforced concrete structures Non-linear Hebbian Learning (NHL), Data-driven NHL, FCM, NB, Multilayer Perceptron (MLP), Bayes net, J48	100 experimental records	Inputs: Column eccentricity, column load, thickness of the cover, water cement ratio, corrosion of reinforcement, shrinkage, spacing of bars, crazing, temperature changes, thermal movement, chloride attack Output: Crack Data-driven NHL is best

(continued)

Table 7.1 (continued)

Author(s) (1)	Application, techniques, and performance measures employed (2)	Case study, data source (3)	Remarks/inferences, (4)
Luo et al. (2022)	Cost of prefabricated building, taking into consideration dynamic interactions FCM, Delphi Method, Content Analysis	Prefabricated building cost in a resettlement house, Nanchang, China	Cause: Scale effect, prefabrication rate, quality of industrial workers, precast concrete production efficiency, transport efficiency, construction management level, standardized and further optimization design Effect: Building cost Nine levels of MF are developed S-shaped curve and hyperbolic tangent functions were utilized as activation functions Higher influencing factors: Scale effect, standardized design, construction management level
Perera et al. (2019)	Structural health monitoring of fibre-reinforced polymer strengthening systems Agglomerative Hierarchical clustering (HC), KMe	Two beams for Electromechanical impedance-based damage detection	Locally damaged areas were identified. Both clustering algorithms provided valuable insights for two beams
Gajan (2021)	Status of shallow foundations in an earthquake loading situation Distance-weighted KNN, Multiple Linear Regression (MLiR) MAPE	140 datasets from the Rocking Foundation database on centrifuges and shaking tables	Inputs (representative): Peak ground acceleration Outputs: Acceleration amplification ratio, total settlement, normalized seismic energy dissipation MLR and KNN perform satisfactorily
Qin et al. (2021)	Prediction of pore-water pressure Long short-term memory (LSTM) MSE, R^2, RMSE, MAPE	1600 datasets from a field monitoring project, Green Hart Tunnel, Netherlands	Inputs: Slurry pressure, advance velocity, drilling, standstill duration, distance along the tunnel axis Output: Pore-water pressure LSTM performed satisfactorily

(continued)

Table 7.1 (continued)

Author(s) (1)	Application, techniques, and performance measures employed (2)	Case study, data source (3)	Remarks/inferences, (4)
Zhang et al. (2021a)	Soil–structure interface modelling Bi-directional Long short-term memory (Bi-LSTM) MAE, MAPE	4127 datasets from the experimental setup	Inputs (representative): Normal stress, shear stress, normal stiffness and displacement, relative density, normalized roughness Outputs: Normal stress, normal displacement, shear stress Bi-LSTM accurately captured behaviour of soil–structure interfaces
Chala & Ray (2023)	Classification of soil using cone penetration test database DT, RF, ANN, Support Vector Machine (SVM) Confusion Matrix, Precision, Sensitivity, Overall Accuracy, F1-score	232 cone penetration test datasets	Inputs: Depth, sleeve friction total, effective vertical stresses, tip resistance, friction ratio Output: Soil classification SVM is best followed by ANN based on overall accuracy
Chen et al. (2023)	Enhancement of sample liquefaction data Wasserstein Generative Adversarial Networks (WGAN), Synthetic Minority Oversampling Technique, SVM, RF, NB, KNN Precision, Accuracy, Recall, F1-score, Spearman correlation coefficient	191 liquefied and 131 non-liquefied cases	Inputs: Earthquake magnitude, sleeve friction ratio, cone tip resistance, maximum horizontal ground surface acceleration, effective and total stress Output: Liquefaction prediction WGAN outperformed the other employed
Asadi (2016)	Prediction of the strength of anisotropic and intact rock masses Mamdani FIS, Genetic Algorithm (GA) R^2, RMSE, Mean Bias Error (MBE)	Hypothetical situation	Inputs: Confining pressure, tensile strength, uniaxial compressive strength Output: Strength of intact rocks GA can select important rules from a preliminary rule base

(continued)

Table 7.1 (continued)

Author(s) (1)	Application, techniques, and performance measures employed (2)	Case study, data source (3)	Remarks/inferences, (4)
Kennedy et al. (2021)	Prediction of two coefficients for treated unsaturated lateritic soil ANFIS, ANFIS optimized by Differential Evolution (DE), GA, Ant Colony Optimization (ACO), MLR, Particle Swarm Optimization (PSO) R^2, RMSE, MAE	121 datasets corresponding to treated specimens	Inputs: Clay activity and content (%), hybrid cement (%), percentage nano-structured quarry fines, fractional angle Output: Coefficients of curvature as well as uniformity ANFIS is the best
Ding et al. (2021)	Prediction of shear strength of soil ANFIS, ANFIS optimized by Henry Gas Solubility Optimization (HGSO), GA, Imperialist Competitive Algorithm (ICA) R^2, RMSE	Hai Phong-Thai Binh Coastal National Road Project, Vietnam	Inputs: Liquid and plastic limits, specific gravity, void ratio, clay, moisture contents Output: Soil shear strength ANFIS, supported by HGSO, is the best
El Shinawi et al. (2021)	Swelling potentiality of soil ANFIS optimized by Ant Lion Optimizer (ALO), Gray Wolf Optimizer (GWO), Reptile Search Algorithm (RSA) R^2, RMSE, MAE	108 soil samples from the experimental setup, El Sherouk City, Egypt,	Inputs: Geotechnical properties, chemical composition Output: Swelling pressure, free swelling index ANFIS supported by RSA is suitable
Shelia & Hoogenboom (2020)	Classification of soil profile data Ward Method of Hierarchical Clustering (HC) Modified distance matrix	10,253 soil profiles, 47,800 horizons from 149 countries	Inputs (representative): Sampling location for each soil profile, soil classification, soil colour, silt, sand, coarse fraction, clay, cation exchange capacity, bulk density, organic carbon, pH in water, KCl, total nitrogen content Output: Clustering of soil profile data

(continued)

Table 7.1 (continued)

Author(s) (1)	Application, techniques, and performance measures employed (2)	Case study, data source (3)	Remarks/inferences, (4)
Seaton et al. (2021)	Soil health cluster analysis Ward Method of HC	1350 topsoils (0–15 cm), Wales	Inputs: Soil surface water repellency, water content, pH, bulk density, carbon concentration, total nitrogen Output: Classification of soils based on topsoil properties Topsoil classification is more associated with the type of land use
Chen et al. (2020)	Landslide susceptibility assessments GA, PSO, SVM, ANN Accuracy, Area Under the Curve-Receiver Operating Characteristic (AUC-ROC)	Database of 335 landslides, Achaia Regional Unit, Northern Peloponnese, Greece	Inputs (representative): Slope angle and aspect, elevation, stream power index, plan curvature, distance to faults and river, hydrological cover, lithology Output: Landslide susceptibility ANN is for feature selection; PSO is for structural parameter optimization of ML models. Both ML techniques performed well. SVM is superior in learning accuracy and AUC-ROC value
Wang et al. (2020)	Detection of bursts in district metering areas RNN, LSTM MAPE, MSE, MAE, True Positive Rate (TPR), Relative Error (RE), Detection accuracy, False Positive Rate (FPR)	Real-life district metering areas in South China	Inputs: Daily time series representing the periodicity of an entire week Output: Flow of present district metering areas for subsequent time steps The chosen methods performed exceptionally well
Madhuri et al. (2021)	Flood susceptibility assessment XGBoost, LR, KNN, SVM, AdaBoost	Hyderabad, India	XGBoost is the best Considered impact of climate change scenarios
Karamoutsou & Psilovikos (2021)	Deep Learning (DL) in water resources Deep Neural Network (DNN) MAE, MSE, Nash–Sutcliffe Efficiency (NSE)	The catchment area of Lake Kastoria, Western Macedonia, Greece	Inputs: Chlorophyll-a, conductivity, pH, turbidity, temperature, ammonia nitrogen, nitrate nitrogen Output: Dissolved oxygen

(continued)

Table 7.1 (continued)

Author(s) (1)	Application, techniques, and performance measures employed (2)	Case study, data source (3)	Remarks/inferences, (4)
Ghimire et al. (2021)	Prediction of Streamflow MLP, CNN-LSTM, DNN, CNN, LSTM, Extreme Learning Machine (ELM), DT, RF, GB Regression, Multivariate Adaptive Regression Splines (MARS), XGBoost R^2, MAE, Relative RMSE, Kling Gupta Efficiency (KGE) , RMSE, absolute percentage bias, integral normalized MSE, Relative Absolute Error (RAE), Normalized Root Mean Square Error (NRMSE), Percentages of MAE, Willmott Index (WI), Percentages of RMSE, NSE, Legate, McCabe Index	Brisbane River and Teewah Creek, Australia	Inputs: Lagged values of streamflow Output: Future streamflow CNN-LSTM performs superior to other models
Guo et al. (2021)	Multi-Step streamflow prediction LSTM, Gated Recurrent Unit (GRU), GWO-least-squares SVM, Bayesian Model Averaging NSE, RMSE, MAE	25 reservoir systems, Zhoushan Islands, China	Inputs: Antecedent precipitation, Antecedent evaporation, antecedent streamflow, forecasted precipitation and evaporation Output: Forecasted streamflow LSTM and GRU accomplished equally well on streamflow forecasts
Asaad et al. (2022)	Streamflow prediction ANN, MLP, LSTM, ANFIS MAE, RMSE, R^2	Kucukmuhsine station, Turkish province of Konya	Inputs: Several delayed streamflows Outputs: Streamflow LSTM had a better performance

(continued)

Table 7.1 (continued)

Author(s) (1)	Application, techniques, and performance measures employed (2)	Case study, data source (3)	Remarks/inferences, (4)
Forghanparast & Mohammadi (2022)	Streamflow prediction CNN, LSTM, Self-Attention LSTM, ELM KGE, R^2	Headwaters of the Colorado River, Texas	Inputs: Precipitation at each month, last month, second last month; evapotranspiration at each month, last month, second last month Output: Streamflows LSTM, Self-Attention–LSTM performed better than the CNN and ELM
Vogeti et al. (2022)	Prediction of Streamflow Bi-LSTM, Wavelet Neural Networks (WNN), XGBoost	Lower Godavari Basin, India	Random search-based hyperparameter tuning XGBoost performed exceptionally well 4 SSPs and 8 decadal time horizons
Aoulmi et al. (2023)	Streamflow prediction CNN, CNN-ICA, CNN-GWO MAE, RMSE, R^2, NSE, WI, Peirce skill, extreme dependency scores	Seybouse Basin, Algeria	Inputs (representative): Total precipitation, surface pressure, Arctic and Antarctic oscillations, Pacific North American pattern, North Atlantic oscillation Output: Streamflow CNN-GWO is superior
Latif & Ahmed (2023)	Streamflow prediction LSTM, Boosted Regression Tree (BRT) RMSE-Standard Deviation ratio, MAE, RMSE, R^2, MSE, NSE	Dokan Dam, Lesser Zab River, Iraq	Inputs: Several delayed streamflows Outputs: Streamflow LSTM significantly outperformed BRT
Mohammed et al. (2023)	Groundwater quality index prediction MLP, SVR MSE, RMSE, MAE, R^2	20 groundwater samples, Northern Khartoum area, Sudan	Inputs: 11 physicochemical parameters Output: Groundwater quality index Both algorithms performed well

(continued)

Table 7.1 (continued)

Author(s) (1)	Application, techniques, and performance measures employed (2)	Case study, data source (3)	Remarks/inferences, (4)
Zaresefat et al. (2023)	Identification of suitable artificial groundwater recharge areas Three-layer ANN MSE, RMSE, R^2	1000 randomly selected locations across Iranshahr Plain, province of Sistan and Baluchistan, southeast Iran	Inputs: Precipitation, geology, soil transmissivity, land use land cover, distance from major rivers, unsaturated zone thickness, slope maps, water quality Output: Suitable artificial groundwater recharge sites
Tirupathi & Shashidhar (2019)	Surface water quality assessment Mamdani FIS	River basins in India, Malaysia, and the United States	Inputs: Biological oxygen demand, electrical conductivity, chemical oxygen demand, dissolved oxygen, faecal coliforms, pH, turbidity, nitrogen, alkalinity, suspended solids Output: Fuzzy regional water quality index Three levels of MF Fuzzy rules for India: 2187 [resulting due to three levels and seven variables] Fuzzy rules for USA: 243 [resulting due to three levels and five variables]
Belvederesi & Thompson (2019)	Predicting the impact of hazardous pipeline accidents ANFIS Adverse impact	814 datasets	Inputs: Commodity transported, location, release mode, maximum operating pressure, ratio of nominal diameter and nominal wall thickness Output: Effect on water, wildlife, and soil contamination
Taylan et al. (2021)	Modelling of air quality Non-linear Autoregressive with exogenous input with ANN, ANFIS, Hybrid data-driven ANN RMSE, MAPE	1771 datasets were collected for pollutants, Jeddah city-industrial zone, Saudi Arabia	Inputs: Sulphur dioxide, nitrogen oxide, carbon monoxide, hydrogen sulphur, ozone, particulate matter Output: Air quality index ANFIS is found to be preferred

(continued)

Table 7.1 (continued)

Author(s) (1)	Application, techniques, and performance measures employed (2)	Case study, data source (3)	Remarks/inferences, (4)
Kikon et al. (2023)	Drought index over an arid region GA-ANFIS, PSO-ANFIS, Generalized Regression Neural Network (GRNN) R^2, NRMSE, NSE, Normalized mean bias, MAE	Rajasthan, India	Inputs: Precipitation Output: Drought index PSO-ANFIS and GA-ANFIS are better than GRNN
Solana-Gutiérrez et al. (2017)	Prediction of river management responses FCM	Esla River, Iberian Peninsula	Causes: Hydroelectric production, socio-economic aspects, urban uses, river connectivity, water quality, agro-forestry production, sediments dynamics, infrastructure, dams, and weirs, natural water flow regime, in-stream communities, riparian landscapes, cross-barriers, bank conditions, and vegetation Effect: River management plan
Pessoa et al. (2018)	Delineation of homogeneous regions for streamflow Fuzzy Cluster Means (FCMe) Pakhira-Bandyopadhyay-Maulik Validity Index	208 stream gauges, Amazon region	10 optimal clusters were identified
Khazaeiathar et al. (2022)	Streamflow modelling FCMe, Autoregressive Moving Average (ARMA) RMSE	Daily discharge of four different river stations, Hesse state, Germany	The performance of the autoregressive moving average was reliable. Prediction is simplified after clustering daily streamflow time series
Cho & Lee (2020)	Allocation of waste load in a river Max–min operator GA	Yeongsan River basin, Korean Peninsula	River water quality and the cost of pollutant load reduction are the objective functions

(continued)

Table 7.1 (continued)

Author(s) (1)	Application, techniques, and performance measures employed (2)	Case study, data source (3)	Remarks/inferences, (4)
Vasan et al. (2021)	Fuzzy optimization in WDN design Self-Adaptive Cuckoo Search Algorithm (CSA)	Hanoi and Pamapur WDN	Three objective functions, three MF The hyperbolic MF is the best
Mirzaie et al. (2021)	Combined use of reclaimed wastewater and groundwater under uncertainty PSO, Multiobjective PSO, Fuzzy Optimization, Non-Fuzzy Optimization	Varamin Plain, Tehran, Iran	Objective functions are maximization of the aquifer recharge, net benefits, and minimization of nitrate leaching Single-objective and multiobjective problems were considered, focusing mainly on cropping patterns Fuzzy models performed better than non-fuzzy models in single- and multiobjective situations
Liu et al. (2021)	Prediction of shared-parking demand CNN-LSTM, MLP, MLR, LSTM, and two more algorithms MAE, RMSE	Shared-parking dataset, Chengdu, China	Inputs: Number of shared-parking vehicles arriving and leaving in a chosen time interval at a region Output: Prediction of shared-parking demand for 1, 1.5, 2 and 3 h CNN-LSTM performed better than the remaining models
Li et al. (2021)	Classification of road crack Combining CNN architecture with K-Means (KMe) Accuracy, Precision, Mean average precision, Recall	Experiments based images	Input: Crack image dataset Output: Road crack classification The fusing of two algorithms is found to be promising
Dai et al. (2022)	Prediction of short-term road surface temperature GRU-LSTM, CNN-LSTM, SVR, Back-Propagation (BP)-ANN MAE, MSE, MAPE	8640 samples, road weather station, China	Inputs: Humidity, temperature, rainfall, wind speed, road surface temperature Output: Forecast of road surface temperature ahead of 1, 3, and 6 h The prediction accuracy of GRU-LSTM is higher than other models

(continued)

Table 7.1 (continued)

Author(s) (1)	Application, techniques, and performance measures employed (2)	Case study, data source (3)	Remarks/inferences, (4)
Li & Abdel-Aty (2022)	Likelihood prediction of real-time crash Temporal Attention-based LSTM, LR, XGBoost, RF, LSTM-CNN, Temporal Attention-based CNN-LSTM AUC-ROC, Sensitivity, False Alarm Rate	6 arterials from Orlando, Florida (includes 110 signalized intersections and 208 road segments) 3,151,169 Lynx records, 138,285 Lytx records, and 162 crashes	Input: Crash and trajectory data Output: Crash likelihood for urban arterials Temporal attention-based CNN-LSTM performed well compared to other employed models
Assi et al. (2022)	Space allocation as well as signal timing problem Dynamic lane assignment optimization model, DNN, KNN, RF, DT Accuracy, F1-score, Error, Precision, False Discovery Rate, Recall, False Negative Rate (FNR), Average processing time	4-leg intersection, Dhahran, Saudi Arabia	Inputs (12 in number): 3 turning movements at a 4-legged signalized intersection Outputs (5 in number): Optimal lane configuration at signalized intersection, Optimal cycle length DL model outperforms others
Morris et al. (2022)	Anomaly detection in traffic data Variational autoencoder-based model, RNN, GRU, LSTM, Liquid Time Constant Network-Neural Circuit Policies, Liquid Time Constant (fully connected wiring) RMSE, Precision, Recall, F1-score	320 continuous count stations locations and 1645 video detection system sites, Georgia Department of Transportation	Inputs: Traffic volume of the training and testing days Output: Anomaly detection in traffic data A fully connected liquid time constant is best due to the lowest prediction error

(continued)

Table 7.1 (continued)

Author(s) (1)	Application, techniques, and performance measures employed (2)	Case study, data source (3)	Remarks/inferences, (4)
Gao et al. (2022)	Modelling of pavement performance CNN, LSTM, CNN-LSTM, Classification And Regression Trees (CART), ANN, RF, MLR R^2	378 condition attributes and 100,000 datapoints from the Pavement Management Information System, Texas Department of Transportation	Inputs: Maintenance work records, pavement condition indicators, road functional class, traffic, pavement type information Output: Pavement deterioration CNN is the best
Prieto et al. (2022)	Prediction of service life of pavements Mamdani FIS	Airport network, Viña del Mar, Central Chile	Inputs: Cracking, potholes, ravelling, temperature, rainfall; five sub-levels of MF Output: Service life of pavement index with nine sub-levels
Ali et al. (2022)	Prediction of pavement condition index Mamdani FIS R^2, RMSE, MAE	Pavement distress data, USA and Canada	Inputs: Fatigue, longitudinal and transverse cracking, rutting, block cracking, bleeding, potholes, patching, ravelling Outputs: Pavement condition index 27 rules were formed Many defuzzification methods were used
Alas et al. (2020)	Modelling of polymer nanocomposite-modified asphalt binder ANFIS, ANN R^2, RMSE	381 datasets from the experimental setup	Inputs: Mechanical test parameters, penetration, softening point, frequency, temperature Output: Complex module Both ANN and ANFIS performed satisfactorily

(continued)

Table 7.1 (continued)

Author(s) (1)	Application, techniques, and performance measures employed (2)	Case study, data source (3)	Remarks/inferences, (4)
Heidaripanah & Hassani (2021)	Prediction of dynamic modulus of asphalt ANFIS-FCMe, ANFIS- Subtractive Clustering (SC) R^2, RMSE, MAE	1320 test results conducted at the University of Maryland	Inputs: Slope and intercept of temperature susceptibility relationships, effective bitumen content, asphalt content, air void content, cumulative passing of the sieves, confining stress, temperature, loading frequency Output: Dynamic modulus Both variations of ANFIS are performing satisfactorily
Zhao et al. (2022)	Urban intelligent transportation FCM	Questionnaire-based survey	Causes: More than 15 causes related to urban intelligent transportation were analyzed Effect: Intelligent transportation construction Highest ranked dimension: Technical support
Bianchini & Bell (2019)	Spatial variability of deflection measurement FCMe	Surveyed 17 flexible pavement sections at various US Army airfields	Inputs: Impulse stiffness modulus, deflection, surface curvature, base damage, base curvature indices Output: Representative basin computed through FCMe
Li et al. (2016)	Characterization of Truck traffic in pavement mechanistic-empirical design KMe, FCMe, Agglomerative HC, Model-based Algorithm	Datasets of 39 weigh-in-motion sites, State of Michigan	Inputs: Factors related to traffic volume adjustment, base traffic volume, and axle load distribution Fuzzy-based method is preferred
Wang et al. (2022a)	Pavement crack segmentation KMe, Gaussian Mixture Model, FCMe, Maximum Entropy Clustering, HC, Mean-Shift Clustering Normalized mutual information, Random index	Sample images that cover typical cracks	FCMe and KMe are sensitive to Gaussian noise

(continued)

Table 7.1 (continued)

Author(s) (1)	Application, techniques, and performance measures employed (2)	Case study, data source (3)	Remarks/inferences, (4)
Huo et al. (2022)	Level-of-service for Bus Rapid Transit FCMe, Simulated Annealing (SA)-Based GA-FCMe Employed seven cluster validity indices	1304 records of Bus Rapid Transit correspond to Guangzhou, Changzhou, and Yichang, China	Inputs: Perceived arrival and waiting times, bus speed, passenger load perceptions, perceived transfer and departure times, overall perception Outputs: Level of service Rapid transit user perception surveys were conducted Simulated Annealing-based GA-FCMe is preferred over FCMe

Table 7.2 Case studies related to Chemical Engineering

Author(s) (1)	Application, techniques, and performance measures employed (2)	Case study, data source (3)	Remarks/inferences, (4)
Fang et al. (2022)	Prediction of chemical toxicity RF, MLR R^2	1792 experimental toxicants towards Tetrahymena pyriformis	Inputs: 9 in number Output: Chemical toxicity RF performed well
Al-Wahaibi et al. (2023)	Fault identification in chemical processes Local Global scale CNN, CNN, CNN-LSTM, multiscale CNN, Global Feature CNN, ANN, Fisher Discriminant Analysis Fault Diagnosis Ratio, Precision, F1-score, TPR, FNR	Fault diagnosis, benchmark Tennessee Eastman process dataset, 52 (includes 11 manipulated variables); 20 simulations	Inputs: The image is transformed from multivariate time-series data Outputs: Fault classification Local Global scale CNN is the best
Theisen et al. (2023)	Digitization of chemical process flow diagrams Pixel-based search algorithm, Faster R-CNN	1005 flowsheets	Inputs: Process flow diagrams Outputs: 47 classes considered for drawing styles of unit operations The proposed approach performs better
Zhang et al. (2023b)	Industrial process fault diagnosis GRU-Enhanced Deep CNN, Deep CNN, GRU Fault Diagnosis Time, FPR, Fault Diagnosis Rate	An acid gas absorption process, Benchmark Tennessee Eastman process	GRU-enhanced deep CNN is the best for two case studies
Xu et al. (2023)	Batteries health state CNN-LSTM-Skip algorithm, CNN-LSTM, one more algorithm RMSE, R^2, MAE	NASA (18,650 lithium-ion batteries), Oxford battery aging datasets (8 Kokam lithium-ion batteries),	Inputs: Current, voltage, sampling time, IC curve, temperature, Base model Output: Battery health state For Oxford datasets: An additional feature is accumulated capacity CNN-LSTM-Skip algorithm performs better than the remaining two

(continued)

Table 7.2 (continued)

Author(s) (1)	Application, techniques, and performance measures employed (2)	Case study, data source (3)	Remarks/inferences, (4)
Nogueira et al. (2023)	Prediction of NOx and CO_2 emissions RF, SHAP, PCA, Factor Analysis, Independent Component Analysis R^2	Experimental testing with different operating conditions Six-cylinder compression-ignition engine with gas natural modes, dual fuel, diesel 40 samples	Inputs: Fuel rail pressure, substitution ratio, air–fuel equivalence ratio, start of injection Outputs: Emissions, Brake thermal efficiency, filter smoke number RF is found to capture the relationship efficiently
Zafari & Ghaemi (2023)	CO_2 capture optimization Radial Basis Function (RBF), ANN, Buckingham Pi theorem, Response Surface Methodology (RSM) MSE, R^2, Average Absolute Relative Error (AARE)	Experimental data	Inputs (representative): Reaction rate constant, amine concentration, CO_2 concentration, total pressure, CO_2 partial pressure, CO_2 diffusion coefficient in gas phase and liquid phase, thickness of gas film Output: Mass transfer flux RBF is the best
Yang et al. (2023)	Graphene oxide membrane optimization GA-BPANN, RF, SVM, BPANN R^2, MAE, MAPE, MSE, RMSE	72 Graphene oxide sheets obtained during 2017–2021	Inputs: Interlayer spacing between the graphene oxide sheets, Operation pressure, roughness and thickness of graphene oxide layer, zeta potential Outputs: Water flux and rejection GA-BPANN is the best
Ge et al. (2023)	Detection of heavy metal pollutants Terahertz spectroscopy, SVM, Deep SVM, DNN Precision, Accuracy, F1-score, Recall	180 experimental datasets	Deep SVM is superior to SVM and DNN

(continued)

Table 7.2 (continued)

Author(s) (1)	Application, techniques, and performance measures employed (2)	Case study, data source (3)	Remarks/inferences, (4)
Zarei et al. (2023)	Removal efficacy of hydrogen Sulphide MLR, SVM R^2, RMSE, Efficacy	Experimental setup	Inputs (representative): Concentration of hydrogen sulphide in the biogas stream input to the biofilter Output: H_2S removal efficiency SVM results are tallying with experimental data
Qian et al. (2023)	Prediction of Urban Gas Consumption CatBoost hybridized with (Phasor PSO, Artificial Bee Colony, Satin Bowerbird algorithm, Battle Royale Optimizer, GWO, Fruit Fly Optimization Algorithm, Urban Gas Consumption) RMSE, MAPE, MAE, RAE, R^2, Normalized MSE	4477 datasets	Inputs: Pressure, price, humidity, temperature, wind speed Output: Urban gas consumption Catboost-Phasor PSO had the best performance
Godwin et al. (2023)	Combustion performance prediction of ethanol-powered spark ignition engine ANN, Ensemble Least Squares Boosting ML techniques R^2, RMSE, MSE, MAE	Experimental datasets	Inputs: Brake-specific fuel consumption, engine load Outputs: Exhaust gas temperature, hydrocarbons, brake thermal efficiency, carbon dioxide, carbon monoxide, nitrogen oxides under different operating situations Ensemble Least squares boosting ML is the best compared to ANN

(continued)

Table 7.2 (continued)

Author(s) (1)	Application, techniques, and performance measures employed (2)	Case study, data source (3)	Remarks/inferences, (4)
Ahmadi et al. (2020)	Flash separator fuzzy dynamic modelling Mamdani FIS MAPE	Knowledge base and experimental setup	Inputs: Feed temperature, feed pressure, outlet gas pressure, outlet liquid pressure, feed-molar fraction Outputs: Gas molar fraction, separator temperature, separator level, separator pressure, liquid molar fraction The proposed number of rules is 5^{52} The linguistic composition variable method is used to decrease rules to 7150
Dubey et al. (2023)	Modelling for Cr (VI) adsorption ANFIS SSE, MSE, RMSE, R^2	Experimental-based	Inputs: Stirring rate and time, initial concentration, contact time, dosage, pH Output: Cr (VI) Adsorption ANFIS can simulate the experimental data
Morone et al. (2021)	Valorization of organic waste flows FCM using ANN	National analysis to understand the Italian waste system Experts interaction	Economic and financial strategies and improvement in collection systems yield positive effects on the outcomes
Liu et al. (2022)	Prediction of ozone Evidential Reasoning (ER)-FCM, Real-coded GA-FCM, PSO-FCM, NHL-FCM, Simple Average model-FCM, Weighted Average model-FCM, Majority Voting model-FCM MAE, MSE, RMSE, Friedman, and Nemenyi tests	Data from Fuyang City and Lanzhou City	Causes: CO, NO_2, SO_2, temperature and humidity Effect: Prediction of O_3 trend ER-FCM achieves relatively better prediction accuracy

(continued)

Table 7.2 (continued)

Author(s) (1)	Application, techniques, and performance measures employed (2)	Case study, data source (3)	Remarks/inferences, (4)
Xue et al. (2018)	Cluster analysis of Japanese pollutant release and transfer register to understand release and toxicity characteristics FCMe Xie and Beni Index, Partition coefficient	Classification of 462 chemicals	Cluster fuzziness is 2 15 chemical features comprise releases and toxicities Classified into 5 clusters FCMe classified effectively
Jafarzade et al. (2023)	Modelling Cadmium in groundwater resources ANFIS-FCMe, ANFIS-SC R^2, Sum of Square Error (SSE), RMSE	51 sampling locations, 158 water samples at Neyshabur city, Central desert of Iran	Inputs: Dissolved solids, electroconductivity, turbidity, pH Output: Cadmium availability ANFIS-FCMe is slightly better than ANFIS-SC
Zhang et al. (2023c)	Prediction of hybrid nanofluids ANFIS-SC, Grid Partitioning, FCMe MSE, MAE, MAPE, R^2, WI	Experimental datasets	Inputs: Density, thermal conductivity ratio, specific heat capacity, dynamic viscosity ratio Output: Thermophysical properties of hybrid nanofluids containing multiwalled carbon nanotubes and oxide nano-sized materials Optimal grid partition-ANFIS is better than other ANFIS approaches in modelling thermal conductivity and specific heat

(continued)

Table 7.2 (continued)

Author(s) (1)	Application, techniques, and performance measures employed (2)	Case study, data source (3)	Remarks/inferences, (4)
Elshenawy et al. (2022)	Fault detection and diagnosis strategy PCA, KNN, PCA-KNN, FCMe, FCMe-KNN Fault detection and alarm rates, Precision, Accuracy	Numerical example: Tennessee Eastman chemical process 600 samples	41 process measurements and 11 manipulated variables are employed 16 faults are recognized as step changes FCMe-KNN approach reduced the computational cost The optimum cluster size was identified based on indicators
Leite et al. (2023)	Adiabatic styrene reactor optimization Generalized DE	Experimental and related studies	Analyzed single and multiobjective (three objectives) Pareto sets were obtained for different reactor configurations The effect of the steam ratio on reactor efficacy was studied
Cortez-González et al. (2023)	Process optimization DE, Aspen one Weighted function and dynamic self-adaptive techniques	Five benchmark functions	Dynamic self-adaptive technique supported by DE is the best
Zhang et al. (2021b)	Application potentiality of multiobjective dynamic DE with parameter self-adaptive strategies Other algorithms that coupled to multiobjective DE are: Non-dominated Sorting Genetic Algorithm (NSGA-II) and GWO, Self-adaptive Mutation Operator, Ranking-based Mutation Operator, Individualized-Instruction TLBO Inverted generational distance, Spread	Tests on 18 numerical experiments and benchmark functions and 3 biochemical processes	The self-adaptive model performed better than the 5 competitors

(continued)

Table 7.2 (continued)

Author(s) (1)	Application, techniques, and performance measures employed (2)	Case study, data source (3)	Remarks/inferences, (4)
Bi et al. (2023)	Discrimination technique of biomass slagging tendency PSO-DNN, RNN, LSTM F1-score, Accuracy, Recall, Precision, Spearman correlation analysis	114 types of biomass obtained from various sources are the datasets	Inputs: Biomass ash content (%) and 13 types of chemical elements Output: Slagging type PSO-DNN is the best
Wang et al. (2022b)	Potentiality of RF, MLP, SVM, PSO, GB for hyperparameter tuning process; NSGA-II-based multiobjective optimization; a number of multicriteria decision-making techniques R^2	Two case studies, combustion process in a power plant and supercritical water gasification process	An integrated framework is helpful

Table 7.3 Case studies related to Mechanical Engineering

Author(s) (1)	Application, techniques, and performance measures employed (2)	Case study, data source (3)	Remarks/inferences, (4)
Zhang & Yin (2021)	Modelling from the image of particles to mechanical characteristics CNN, Bi-LSTM MAE, MAPE, MSE	200 datasets of biaxial tests generated using the 2-dimensional discrete element method; 600 images	CNN precisely acquired the particle information Bi-LSTM captured impacts of relative density and particle morphology on global mechanical behaviour of granular materials
Wan et al. (2023)	Prediction of power load LSTM, CNN-LSTM, CNN-LSTM-Attention-based mechanism Pearson Correlation Coefficient (PCC), MAE, MAPE, RMSE	Two thermal power (stream turbine) units; Daily operation data of 250 dimensions from a steam turbine unit, Zhejiang Province, China	Inputs: Temperature, pressure, flow-related features Output: Short-term power load CNN-LSTM-Attention mechanism is the best
Abbaskhah et al. (2023)	Horizontal axis wind turbine optimization CNN, MLP MSE, MAE, SSE, R^2	Data from numerical simulation	Inputs: Pitch angle, rotation speed, wind speed, dimpled or original blades Output: Torque, thrust MLP and CNN are performing better
Ranawat et al. (2023)	Blockage detection in centrifugal pump Bi-LSTM, LSTM, SVM, SVM–Grid Search Optimization, SVM– Bayesian Optimization, XGBoost F1-score, Accuracy, Recall, Precision	The experimental facility, 10 different pump conditions; 5000 samples	LSTM performance is superior to Bi-LSTM and others in terms of accuracy
Tian et al. (2023)	The remaining effective life of turbofan engine prediction Spatial correlation and temporal attention-based LSTM, number of related algorithms (variations) RMSE, Score	Datasets correspond to two different turbofan engine simulations	Spatial correlation and temporal attention-based LSTM performance is the best

(continued)

Table 7.3 (continued)

Author(s) (1)	Application, techniques, and performance measures employed (2)	Case study, data source (3)	Remarks/inferences, (4)
Yuan et al. (2023)	Prediction of mechanical behaviours of nitrile butadiene rubber materials Self-adaptive PSO-ANN, ANN-Levenberg–Marquardt, ANN-Gradient Descent, PSO-ANN, WGAN MSE, R^2, RMSE	Experimental setup	Input: Stress Output: Strain Stress–strain relationships of rubber materials with various hardness and loading rates are studied Self-adaptive PSO-ANN is the best
Liu et al. (2023)	Composition design of high-performance copper alloy ANN-GA MSE	Experimental setup	Inputs: Number of alloys Output: Performance of copper alloys ANN-GA demonstrated good agreement with experimental output
Zhou et al. (2023)	Fault diagnosis Fuzzy regular least squares twin SVM extended to a multiclassification algorithm	Data of bearing fault diagnosis	Fuzzy regular least squares twin SVM algorithm has good generalization and anti-outlier ability; it also has higher reliability of fault diagnosis
Ding et al. (2023)	Fuel cell air compressor performance prediction BPANN optimized by GA and SVM, GA-BPANN-SVM MAE, MAPE, RMSE, R^2	264 Experimental datasets	Inputs: Corrected speed and pressure ratio Output: Corrected flow rate GA-BPANN-SVM showed a good performance than others
Gao et al. (2023)	Noise recognition of moving parts in the sealed cavity Fused ML-based on CatBoost, XGBoost, LR; XGBoost, CatBoost, RF, DT, LR, XGBoost-LR, CatBoost-LR Accuracy, Recall, Precision, F1-score, AUC-ROC	5 datasets from UCI public datasets (266,196 samples), 5 noise datasets (14,731 samples) from PIND devices	The accuracy of the fused technique is higher than that of traditional stacking
Xiang et al. (2024)	Prediction of metal tubes bending performance Parameters—weight-adaptive CNN, VGG, ResNet, Densenet; RF for ranking the input parameters; ABAQUS, Latin Hypercube Sampling R^2, MAPE, WI, NSE	Datasets of 6061 aluminium tubes, experimental verification	Inputs: Tube process and geometric parameters related to pressure die Outputs (representative): Short-axis variation rate, wall-thickening, thinning ratios Parameters—weight-adaptive-CNN performed best

(continued)

Table 7.3 (continued)

Author(s) (1)	Application, techniques, and performance measures employed (2)	Case study, data source (3)	Remarks/inferences, (4)
Çağıl et al. (2023)	Prediction of vibration of a diesel engine MLR, ANFIS R^2, MSE, MAE, RMSE	2074 experimental datasets	Inputs: x-axis and y-axis (m/s^2), NH_3 additive rate, engine speed (m/s), RMS (m/s^2) Output: Vibration magnitude computed with z-axis (m/s^2) ANFIS is the best
Sundar & Mewada (2023)	Thermal performance factor of nanofluids MLP-ANN, ANFIS RMSE, MSE, R^2	Experimental setup	Inputs: Volume of concentration, Reynolds number Output: Thermal and frictional entropies, thermal performance factor, energy efficiency ANFIS is the best
Zare et al. (2022)	Wind energy deployment pathways FCM-based approach	Wind energy deployment in Iran Data mainly through surveys	26 criteria belong to 6 groups, i.e., economic, political, technological, social, legal, and environmental. Interlinkages between them are also discussed 4 possible scenarios 10 logistic terms to define positive and negative relationships among criteria
Pereira et al. (2020)	Impacts of energy-change FCM—System Dynamic Approach	Portugal	Analyzed the role of FCM and system dynamic approach
Tung et al. (2023)	Voronoi structures optimization GA, Finite Element Simulation	Experimental setup	Voronoi structures optimized by the GA enhanced strength, stiffness, and toughness values by $\sim$30%
Song et al. (2023)	Layout of wind farm Adaptive Granularity Learning Distributed PSO, Yawed Gaussian Wake, and Wake Merging Models	Various data sources to replicate realistic wind farm	Adaptive granularity learning distributed PSO solves the problem effectively

Table 7.4 Case studies related to Electronics and Computer Science Engineering

Author(s) (1)	Application, techniques, and performance measures employed (2)	Case study, data source (3)	Remarks/inferences, (4)
Nayak et al. (2021)	Hand gesture recognition Memetic Firefly technique with LGBoost, NB, Linear Regression (LiR), Stochastic Gradient Descent, Linear and Quadratic Discriminant Analyzer, RF, DT, KNN, LGBoost, GB, Adaboost, Firefly Technique with LGBoost Precision, Accuracy, F1-score, Recall, AUC-ROC	5 hand postures of 12 users	The Memetic Firefly technique with LGBoost is superior to that of others
Takahashi et al. (2022)	Data supplement for brain–computer interface system Electroencephalogram, CNN-LSTM, Common Spatial Pattern, Fully connected ANN with features, extraction, Empirical mode decomposition	200 recordings for each subject	Created 300 artificial data from 60 real-data CNN-LSTM and an empirical mode decomposition process improved electroencephalogram pattern recognition
Kilincer et al. (2022)	Comprehensive intrusion detection environment GB, LGBoost, XGBoost, Catboost, AdaBoost, KNN, NB, MLP, DT, Extra Tree Algorithm Accuracy, Recall, Precision, F1-score	Laboratory-based Detection datasets	LGBoost is the best
Koşar & Barshan (2023)	Activity recognition LSTM, 1D and 2D CNN, 1D and 2D CNN-LSTM	Daily and sports activities, UCIHAR dataset	The present study is compared with other published articles. 2D-CNN-LSTM is superior to other employed techniques
Rajeshkumar et al. (2023)	Smart office automation Faster R-CNN-based face recognition with IoT, VGG-16, SVM, Deep CNN, and two more algorithms Accuracy, Specificity, Sensitivity	8421 face images in RGB format	The accuracy range of Faster R-CNN is superior

(continued)

Table 7.4 (continued)

Author(s) (1)	Application, techniques, and performance measures employed (2)	Case study, data source (3)	Remarks/inferences, (4)
Tan et al. (2023)	Microarchitecture-level fault injection environment Saca-FI, CIFAR-10 CNN, LeNet-5, VGG-16 Reliability-associated architectural vulnerability factor	Simulation analysis: CIFAR-10 CNN with Cifar-10 dataset, LeNet-5 with MNIST dataset, VGG-16 with ILSVRC-2012 dataset	Saca-FI mechanism helps assess vulnerability aspects and construct reliable systolic array-based CNN accelerators
Khan et al. (2023)	Malware detection Squeezed-Boosted Boundary-Region Split-Transform-CNN, MLP, SVM, AdaBoost Accuracy, Precision, Sensitivity, Mathews Correlation Coefficient, F1-score, AUC-ROC	IoT Malware dataset Total images 3959 (Benign ware 2486 and Malware 1473)	Squeezed-Boosted Boundary-Region Split-Transform-CNN is performing well than others
Menaka & Samraj (2023)	Edge system recommendation for cloud service providers Hybrid CNN-LSTM	Kaggle online web-based repository	Hybrid CNN-LSTM captured the process effectively
Adedeji (2023)	Energy consumption prediction Multifunctional ANN, Multioutput inverse function ANN MSE, MAE, RMSE	Battery electric vehicle as a case study	An extensive survey on electric vehicles is also part of the paper 9 inputs and 9 outputs The accuracy of multifunctional ANN is greater than that of multi-output inverse function ANN
Muruganandam et al. (2023)	Prediction of K-barriers for intrusion detection DL-based Feed-Forward ANN, GRNN, RF, and one more algorithm, Monte Carlo Simulation, Binary Sensing Model, Mersenne Twister random number generator RMSE, R^2	Synthetic data using Monte Carlo simulation and related approaches	DL-based feed-forward ANN is better than the remaining techniques that were applied

(continued)

Table 7.4 (continued)

Author(s) (1)	Application, techniques, and performance measures employed (2)	Case study, data source (3)	Remarks/inferences, (4)
Cao et al. (2023)	Prediction of power load Improved XGBoost with a Random Grid Search and Windowed Mechanism, LGBoost, XGBoost, RF, LGBoost-XGBoost, CNN-LSTM Symmetric MAPE, MAE, PCC, RMSE, MAPE, Median absolute error, Cosine similarity, One-Way ANOVA	Electricity Load Diagrams based on data of 370 sub-stations from January 1, 2011-January 1, 2015, Portugal EMC, UKDALE, REFIT datasets	Improved XGBoost with a Random Grid Search and Windowed Mechanism predicted short-term power load effectively
Anbarasu et al. (2020)	Maximum power point tracking of the grid-integrated solar system ANFIS-based Fractional Order Proportional Integral Derivative Controller Percentage error	Intensive and operative data developed based on the voltaic cell model	Improved ANFIS-based controller is more efficient than regularly employed controllers
Dong et al. (2021)	Recognition of eye movement Soft multifunctional electronic skin, ANFIS PCA	Soft multifunctional electronic skin mechanism gathered the Electrooculogram, temperature, and hydration data, 200 datasets	Inputs: Electrooculogram, sweat signals, skin temperature Output: Eye movement Integrating ANFIS with a soft, multifunctional electronic skin mechanism can solve eye movement tracking problems
Chen et al. (2022a)	Ventricular arrhythmia classification 3-dimensional phase space diagram, FCMe F1-score, Sensitivity, Accuracy, Specificity, FPR, Discovery, Negative rates, Positive and Negative Predictive Values	32 healthy subjects from the PTB Diagnostic Database; 32 arrhythmic subjects from the CU Ventricular Tachyarrhythmia database	FCMe facilitates the prediction of a prospective arrhythmia before it happens, and its category

(continued)

Table 7.4 (continued)

Author(s) (1)	Application, techniques, and performance measures employed (2)	Case study, data source (3)	Remarks/inferences, (4)
Kumar et al. (2023)	Arrhythmia detection from electrocardiogram signals Coupled FCMe and DNN, RF, LR, KMe, Gaussian Naive Bayes, KNN, SVM, DT, CNN Feature Extractor Recall, Precision, F1-score, Accuracy	Benchmark datasets	Coupled FCMe and DNN are superior compared to other employed techniques for arrhythmia detection
Tyagi & Jha (2023)	Wireless sensor network FCMe-based indices Energy-centric reputation index, internodal distance, relevance index method, degree of the node, distance to the sink	Simulation studies	27 rules are formulated
Ma et al. (2023)	Complex contact phenomena Self-optimized ANN, GA-Sequential Quadratic Programming-ANN, GA-ANN MSE	Simulation studies	GA-sequential quadratic programming-ANN is better than the remaining models
Chandra et al. (2023)	Higher ensured life span of IoT in 5G network GA, Improved GA-Binary ACO, Improved GA- Fast Non-Dominated Sorting Network Simulator-2 Remaining nodes' longevity, Computing time, Energy efficiency	Experimental setup	Improved GA-fast non-dominated sorting achieves a higher lifetime and lower computation time
Zhuang et al. (2023)	Cooperative spectrum sensing Siegel distance-based Fusion Strategy, DE-cooperative Spectrum Sensing, Symmetrized Kullback–Leibler divergence-based DE	Simulation modelling	Symmetrized Kullback–Leibler divergence-based DE is the best
Narayanan et al. (2023)	K-barrier count intrusion detection system PSO-ANN, existing DT, NB, Monte Carlo Simulation Number of invasions, R^2, RMSE, Accuracy, Precision	Monte Carlo Simulation-based synthetic datasets, University of California ML repository	PSO-ANN produced 90% intrusion detection accuracy

Table 7.5 Case studies related to management and allied fields

Author(s) (1)	Application, techniques, and performance measures employed (2)	Case study, data source (3)	Remarks/inferences, (4)
Al-azazi & Ghurab (2023)	Early student performance prediction ANN-LSTM, GRU, RNN, Deep Feed-Forward ANN, RF, and one more algorithm F1-score, Accuracy, Recall, Precision	32,593 students	ANN-LSTM is the best
Xu & Sun (2023)	Prediction of academic performance RF, SVM, KNN Accuracy, F1-score, AUC	432 fifth-grade students	Inputs: Muscle strength, aerobic endurance, body mass index, speed, flexibility Outputs: Academic performance Prediction accuracies of the ML techniques are the same in training and testing
Tchupo & Macht (2022)	Team communication FCM 37 different FCM analysis indicators	Two fuzzy cognitive maps in team communication domain	FCM for team communication is beneficial
Oqaidi et al. (2022)	Students' Performance ML techniques and FCM	Moroccan Higher Education institutions	ML for large datasets. FCM for the limited number of student programs

(continued)

Table 7.5 (continued)

Author(s) (1)	Application, techniques, and performance measures employed (2)	Case study, data source (3)	Remarks/inferences, (4)
Han (2023)	Prediction of mental health in terms of psychological fitness and performance detection for university students Heuristic FCMe, AdaBoost, Data Mining Techniques, Cognitive Trait Model, Felder-Silverman Learning Style, DL-based Mental Health Monitoring system	Kaggle stress dataset	DL-based Mental Health Monitoring Scheme achieved ratios of cognitive development, high student performance, student engagement, and prediction
Vetrimani et al. (2023)	Croup cough image classification problem CNN-GA, deep convolutional WGAN to develop artificial images due to insufficient images F1-score, Accuracy, Precision, Recall	987 X-ray images of paediatric, binary classification of croup cough images	CNN-GA is suitable for weight optimization and classification of images
Jain et al. (2023)	Heart disease prediction Levy flight-based Sunflower Optimization Algorithm, levy flight-CNN, Big Data, NB, ANN, DT, RF, SVM, and Pre-trained CNN Recognition error, accuracy, specificity, sensitivity, processing time, Mathews correlation coefficient	Angiography of 500 patients	Levy flight-CNN is superior to previous studies

(continued)

Table 7.5 (continued)

Author(s) (1)	Application, techniques, and performance measures employed (2)	Case study, data source (3)	Remarks/inferences, (4)
Martinez-Gil & Chaves-Gonzalez (2022)	Interpretable ontology meta-matching Mamdani FIS Precision, recall, and f-measure, Ontology Matches: Levenshtein Similarity, Cosine Similarity, Ratcliff–Obershelp, UMLS Similarity, Q-Gram Similarity	Data reported from the previous studies	20 rules Mamdani FIS approach is effective in interpretable ontology meta-matching compared to the previous studies
Sweidan et al. (2023)	Liver fibrosis diagnosis through Hepatitis C virus Fuzzy-based ontology system works on semantic rule-based techniques, Mamdani FIS F1-score, Recall, Precision, Accuracy	Dataset of real-fibrosis cases Tested on 47 chronic hepatitis C virus cases	A fuzzy-based ontology system is best compared to standard Mamdani FIS in terms of accuracy 74 fuzzy rules were generated
Fatima et al. (2022)	Prediction of operational variables for doctors ANFIS	Questionnaire, 48 valid responses	Doctors using Industry 4.0 had better accuracy during their diagnoses and surgeries than others who are not utilizing said technology Number of MF and 81 rules
Akinnuwesi et al. (2020)	FCM enabled DSS to diagnose rheumatic-musculoskeletal disease Mamdani FIS, FCM, Quadratic, and Medium Gaussian SVM, Ensemble (bagged tree), KNN weighted, Tree (fine, medium, and coarse) Accuracy, Sensitivity, Specificity, AUC	Mainly interviews in Nigeria	Input: 28 variables Output: Rheumatic-musculoskeletal disease diagnosis FCM facilitates the achievement of speed diagnosis, specificity, accuracy, sensitivity The Centre of gravity is the defuzzification method

(continued)

Table 7.5 (continued)

Author(s) (1)	Application, techniques, and performance measures employed (2)	Case study, data source (3)	Remarks/inferences, (4)
Sasan et al. (2019)	Formwork labour productivity GRNN, BPANN, RBF, ANFIS MSE, R^2	221 datasets from 2 high-rise buildings	BPANN performs better
Park et al. (2022)	Stock market prediction Integrating LSTM and the RF framework with multitasking (LFM), RF, LSTM base, Integrating LSTM and the RF framework with Singletasking (LFS), and other algorithms MAPE, RMSE, MAE, Accuracy	Publicly available data	LFM is a suitable model compared to other models
Bhandari et al. (2022)	Stock market index prediction LSTM (single layer and multilayer) RMSE, MAPE, R^2	15 years (2006–2020) of publicly available data	Inputs (representative): Open price, close price, civilian unemployment rate, consumer sentiment index Output: Next-day closing price Single-layer LSTM is better than multilayer LSTM
Wang et al. (2023a)	Stock prediction Interpretable Intuitionistic Fuzzy Inference Algorithm comprising of DNN, XGBoost, Stacking, LGBoost	Experimental dataset	21 stock market-related features used for analyzation The interpretable intuitionistic fuzzy inference algorithm comprising DNN improves the profit by more than 20% compared to other methods

(continued)

Table 7.5 (continued)

Author(s) (1)	Application, techniques, and performance measures employed (2)	Case study, data source (3)	Remarks/inferences, (4)
Boyacioglu & Avci (2010)	Stock market return prediction ANFIS RMSE, R^2, Coefficient of variation	Istanbul Stock Exchange 228 pairs of observations	ANFIS improved stock price prediction
Islam et al. (2023)	Efficient disaster (flood) management CNN-sorting algorithm for image classification, DenseNet-121, Inception v3-based CNN models F1-score, Accuracy, Recall, Precision	Data of flood images	CNN-sorting algorithm is the best
Mehryar & Surminski (2022)	Perceptions of Flood Resilience FCM, Flood Resilience Measurement for Communities	Lowestoft town, England	Integrated FCM has 40 homogenized indicators, i.e., 40 nodes and 161 connections among the nodes

7.2 Further Research Work

No algorithm can be universally employed for all real-world problems. An in-depth understanding of problem-specific knowledge may help to achieve faster and more efficient solutions to a real-world planning problem.

The following potential research areas are identified based on the extensive studies by the authors (including critical analysis of chapters 2–6 in this book). They can be implemented in any domain of engineering, science, and management, which are as follows:

1. Study the applicability of existing AI algorithms to various domains to ascertain their potentiality.
2. Developing new algorithms that efficiently handle non-linear, non-convex, non-differentiable, and multi-modal functions and comparing them with the existing algorithms is a potential research area.
3. Many researchers have developed several evolutionary algorithms. However, few algorithms are only frequently applied to a specific domain. In addition, hybridization of multiple algorithms is another potential area of research.
4. Extending AI and EA into the fuzzy-based uncertain framework is a promising research area.
5. Identifying optimal parameter values in AI algorithms is a significant concern and has enormous potential. EA plays a major role in this direction.
6. Many researchers employed ANFIS, which is partially established on ANN philosophy. The application of DL in the ANFIS framework is a promising research area that can be explored.
7. FCM can be trained with more algorithms, such as HL and EA, to improve its applicability to real-world situations.
8. Hybridization of Blockchain, FIS, and EA is another research area highly suitable for societal-related challenges.

Representative books and journals related to AI are available in Appendix B and will benefit readers of the book for enhanced understanding on the topic.

Revision Questions and Exercise Problems

7.1 Compile case studies related to FIS, ANFIS, and FCM and provide salient conclusions.

7.2 What are the salient observations made while studying case studies related to ML?

Advance Review Questions

7.3 Analyze research papers published in domains mentioned in this chapter related to ANFIS, FIS, and FCM. Present five research areas that are suitable for further work.

7.4 Discuss six recently published research papers on ML and their fuzzy extensions in chemical, mechanical, and computer science domains. What are the techniques applied, and how was hyperparameter tuning made for the parameters?

References

Abbaskhah, A., Sedighi, H., Akbarzadeh, P., & Salavatipour, A. (2023). Optimization of horizontal axis wind turbine performance with the dimpled blades by using CNN and MLP models. *Ocean Engineering, 276*, 114185.

Adedeji, B. P. (2023). Electric vehicles survey and a multifunctional artificial neural network for predicting energy consumption in all-electric vehicles. *Results in Engineering, 19*, 101283.

Ahmadi, M. H. E., Royaee, S. J., Tayyebi, S., & Boozarjomehry, R. B. (2020). A new insight into implementing mamdani fuzzy inference system for dynamic process modeling: Application on flash separator fuzzy dynamic modeling. *Engineering Applications of Artificial Intelligence, 90*, 103485.

Akinnuwesi, B. A., Adegbite, B. A., Adelowo, F., Ima-Edomwonyi, U., Fashoto, G., & Amumeji, O. T. (2020). Decision support system for diagnosing rheumatic-musculoskeletal disease using fuzzy cognitive map technique. *Informatics in Medicine Unlocked, 18*, 100279.

Al-Azazi, F. A., & Ghurab, M. (2023). ANN-LSTM: A deep learning model for early student performance prediction in MOOC. *Heliyon, 9*, e15382.

Al-Wahaibi, S. S. S., Abiola, S., Chowdhury, M. A., & Lu, Q. (2023). Improving convolutional neural networks for fault diagnosis in chemical processes by incorporating global correlations. *Computers & Chemical Engineering, 176*, 108289.

Alas, M., Ali, S. I. A., Abdulhadi, Y., & Abba, S. I. (2020). Experimental evaluation and modeling of polymer nanocomposite modified asphalt binder using ANN and ANFIS. *Journal of Materials in Civil Engineering, 32*, 04020305.

Ali, A., Heneash, U., Hussein, A., & Eskebi, M. (2022). Predicting pavement condition index using fuzzy logic technique. *Infrastructures, 7*, 91.

Anbarasu, E., Muthu, V. P. S., & Basha, A. R. (2020). An Improved power conditioning system for grid integration of solar power using ANFIS based FOPID controller. *Microprocessors and Microsystems, 74*, 103030.

Aoulmi, Y., Marouf, N., Rasouli, K., & Panahi, M. (2023). Runoff predictions in a semiarid watershed by convolutional neural networks improved with metaheuristic algorithms and forced with reanalysis and climate data. *Journal of Hydrologic Engineering, 28*, 04023018.

Asaad, M. N., Eryürük, Ş, & Eryürük, K. (2022). Forecasting of streamflow and comparison of artificial intelligence methods: A case study for meram stream in Konya, Turkey. *Sustainability, 14*, 6319.

Asadi, M. (2016). Optimized mamdani fuzzy models for predicting the strength of intact rocks and anisotropic rock masses. *Journal of Rock Mechanics and Geotechnical Engineering, 8*, 218–224.

Assi, K., Ratrout, N., Nemer, I., Rahman, S. M., & Jama, A. (2022). Framework of big data and deep learning for simultaneously solving space allocation and signal timing problem, *Journal of Transportation Engineering. Part A: Systems, 149*, 04022126.

Belvederesi, C., & Thompson, M. S. (2019). Predicting environmental impact of hazardous liquid pipeline accidents: Application of intelligent systems. *Journal of Environmental Engineering, 146*, 04019104.

Bhandari, H. N., Rimal, B., Pokhrel, N. R., Rimal, R., Dahal, K. R., & Khatri, R. K. C. (2022). Predicting stock market index using LSTM. *Machine Learning with Applications, 9*, 100320.

Bi, Y., Chen, C., Huang, X., Wang, H., & Wei, G. (2023). Discrimination method of biomass slagging tendency based on particle swarm optimization deep neural network (DNN),. *Energy*, 262 (Part A), 125368.

Bianchini, A., & Bell, H. P. (2019). Fuzzy cluster approach for area FWD representative basin from deflection measurement spatial variability. *International Journal of Pavement Engineering, 20*, 844–852.

Boyacioglu, M. A., & Avci, D. (2010). An adaptive network-based fuzzy inference system (ANFIS) for the prediction of stock market return: The case of the istanbul stock exchange. *Expert Systems with Applications, 37*, 7908–7912.

Çağıl, G., Güler, S. N., Ünlü, A., Böyükdibi, Ö., & Tüccar, G. (2023). Comparative analysis of multiple linear regression (mlr) and adaptive network-based fuzzy inference systems (ANFIS) methods for vibration prediction of a diesel engine containing NH_3 additive. *Fuel, 350*, 128686.

Cao, W., Liu, Y., Mei, H., Shang, H., Yu, Y. (2023). Short-term district power load self-prediction based on improved xgboost model. *Engineering Applications of Artificial Intelligence*, 126 (Part A), 106826.

Chala, A. T., & Ray, R. (2023). Assessing the performance of machine learning algorithms for soil classification using cone penetration test data. *Applied Sciences, 13*, 5758.

Chandra, B. R., Kumar, K., Roy, A., Qamar, S., Rahman, M. I., & Saif, A. G. F. (2023). Genetic algorithm for higher ensured lifespan of internet of things in 5G network. *Computers and Electrical Engineering, 106*, 108563.

Chen, H., Das, S., Morgan, J. M., & Maharatna, K. (2022a). Prediction and classification of ventricular arrhythmia based on phase-space reconstruction and fuzzy C-means clustering. *Computers in Biology and Medicine, 142*, 105180.

Chen, M., Kang, X., & Ma, X. (2023). Deep learning-based enhancement of small sample liquefaction data. *International Journal of Geomechanics, 23*, 04023140.

Chen, S. Z., Feng, D. C., Wang, W. J., & Taciroglu, E. (2022b). Probabilistic machine-learning methods for performance prediction of structure and infrastructures through natural gradient boosting. *Journal of Structural Engineering, 148*, 04022096.

Chen, W., Chen, Y., Tsangaratos, P., Ilia, I., & Wang, X. (2020). Combining evolutionary algorithms and machine learning models in landslide susceptibility assessments. *Remote Sensing, 12*, 3854.

Cho, J. H., & Lee, J. H. (2020). Fuzzy optimization model for waste load allocation in a river with total maximum daily load (TMDL) planning. *Water, 12*, 2618.

Cortez-González, J., Hernández-Aguirre, A., Murrieta-Dueñas, R., Gutiérrez-Guerra, R., Hernández, S., & Segovia-Hernández, J. G. (2023). Process optimization using a dynamic self-adaptive constraint handling technique coupled to a differential evolution algorithm. *Chemical Engineering Research and Design, 189*, 98–116.

Dai, B., Yang, W., Ji, X., Zhu, F., Fang, R., & Zhou, L. (2022). An ensemble deep learning model for short-term road surface temperature prediction, *Journal of Transportation Engineering. Part B: Pavements, 149*, 04022067.

Diker, F., & Erkan, L. (2022). The fuzzy logic method in assessing window design for the visual comfort of classrooms at the early design stage. *Journal of Architectural Engineering, 28*, 04022013.

Ding, C., Xia, Y., Yuan, Z., Yang, H., Fu, J., & Chen, Z. (2023). Performance prediction for a fuel cell air compressor based on the combination of backpropagation neural network optimized by genetic algorithm (GA-BP) and support vector machine (SVM) algorithms. *Thermal Science and Engineering Progress, 44*, 102070.

Ding, W., Nguyen, M. D., Mohammed, A. S., Armaghani, D. J., Hasanipanah, M., Bui, L. V., & Pham, B. T. (2021). A new development of ANFIS-based henry gas solubility optimization technique for prediction of soil shear strength. *Transportation Geotechnics, 29*, 100579.

Dong, W., Yang, L., Gravina, R., & Fortino, G. (2021). ANFIS fusion algorithm for eye movement recognition via soft multifunctional electronic skin. *Information Fusion, 71*, 99–108.

Dubey, M., Joshi, M., Thota, C., & Kumar, R. (2023). Novel Template free synthesis of high surface area mesoporous ceria and adaptive neuro fuzzy interference system (ANFIS) modelling for Cr (VI) adsorption. *Materials Science and Engineering: B, 298*, 116819.

Elshenawy, L. M., Chakour, C., & Mahmoud, T. A. (2022). Fault detection and diagnosis strategy based on k-nearest neighbors and fuzzy c-means clustering algorithm for industrial processes. *Journal of the Franklin Institute, 359*, 7115–7139.

El Shinawi, A., Ibrahim, R. A., Abualigah, L., Zelenakova, M., & Abd Elaziz, M. (2021). Enhanced adaptive neuro-fuzzy inference system using reptile search algorithm for relating swelling potentiality using index geotechnical properties: A case study at el sherouk city, egypt. *Mathematics, 9*, 3295.

Fang, Z., Yu, X., & Zeng, Q. (2022). Random forest algorithm-based accurate prediction of chemical toxicity to tetrahymena pyriformis. *Toxicology, 480*, 153325.

Fatima, M., Sherwani, N. U. K., Khan, S., & Khan, M. Z. (2022). Assessing and predicting operation variables for doctors employing industry 4.0 in health care industry using an adaptive neuro-fuzzy inference system (ANFIS) approach. *Sustainable Operations and Computers, 3*, 286–295.

Feng, D. C., Wang, W. J., Mangalathu, S., & Taciroglu, E. (2021). Interpretable XGBoost-SHAP machine-learning model for shear strength prediction of squat RC walls. *Journal of Structural Engineering, 47*, 04021173.

Forghanparast, F., & Mohammadi, G. (2022). Using deep learning algorithms for intermittent streamflow prediction in the headwaters of the Colorado River, Texas. *Water, 14*, 2972.

Gajan. (2021). Application of machine learning algorithms to performance prediction of rocking shallow foundations during earthquake loading. *Soil Dynamics and Earthquake Engineering, 151*, 106965.

Gao, L., Han, Z., & Chen, Y. (2022). Deep learning-based pavement performance modeling using multiple distress indicators and road work history, *Journal of Transportation Engineering. Part B: Pavements, 149*, 04022061.

Gao, Y., Li, C., Sun, Z., & Wang, G. (2023). Fused machine learning based on boosting algorithm for noise recognition of moving parts in sealed cavity. *Measurement, 221*, 113415.

Ge, H., Ji, X., Lu, X., Lv, M., Jiang, Y., Jia, Z., & Zhang, Y. (2023). Identification of heavy metal pollutants in wheat by Thz spectroscopy and deep support vector machine. *Spectrochimica Acta Part A: Molecular and Biomolecular Spectroscopy, 303*, 123206.

Ghimire, S., Yaseen, Z. M., Farooque, A. A., Deo, R. C., Zhang, J., & Tao, X. (2021). Streamflow prediction using an integrated methodology based on convolutional neural network and long short-term memory networks. *Scientific Reports, 11*, 17497.

Godwin, D. J., Varuvel, E. G., & Martin, M. L. J. (2023). Prediction of combustion, performance, and emission parameters of ethanol powered spark ignition engine using ensemble least squares boosting machine learning algorithms. *Journal of Cleaner Production, 421*, 138401.

Guo, Y., Yu, X., Xu, Y. P., Chen, H., Gu, H., & Xie, J. (2021). AI-based techniques for multi-step streamflow forecasts: Application for multiobjective reservoir operation optimization and performance assessment. *Hydrology and Earth System Sciences, 25*, 5951–5979.

Han, H. (2023). Fuzzy clustering algorithm for university students' psychological fitness and performance detection. *Heliyon, 9*, e18550.

Heidaripanah, A., & Hassani, A. (2021). Adaptive neuro-fuzzy inference system to predict the dynamic modulus of Hot Mix Asphalt, *Journal of Transportation Engineering. Part B: Pavements, 147*, 04021043.

Huo, Y., Zhao, J., Li, X., & Guo, C. (2022). Using fuzzy clustering of user perception to determine the number of level-of-service categories for bus rapid transit. *Journal of Public Transportation, 24*, 100017.

Islam, Md. A., Rashid, S. I., Hossain, N. U. I., Fleming, R., Sokolov, A. (2023). An integrated convolutional neural network and sorting algorithm for image classification for efficient flood disaster management. *Decision Analytics Journal, 7*, 100225

Jafarzade, N., Kisi, O., Yousefi, M., Baziar, M., Oskoei, V., Marufi, N., & Mohammadi, A. A. (2023). Viability of two adaptive fuzzy systems based on fuzzy c means and subtractive clustering methods for modeling cadmium in groundwater resource(s). *Heliyon, 9*, e18415.

Jain, A., Rao, A. C. S., Jain, P. K., & Hu, Y. C. (2023). Optimized levy flight model for heart disease prediction using CNN framework in big data application. *Expert Systems with Applications, 223*, 119859.

Karamoutsou, L., & Psilovikos, A. (2021). Deep learning in water resource(s) management: The case study of Kastoria Lake in Greece. *Water, 13*, 3364.

Kennedy, C. O., Shakeri, J., Amini-Khoshalann, H., Salahudeen, A. B., Arinze, E. E., & Ugwu, H. U. (2021). Application of ANFIS hybrids to predict coefficients of curvature and uniformity of treated unsaturated lateritic soil for sustainable earthworks. *Cleaner Materials, 1*, 100005.

Khan, S. H., Alahmadi, T. J., Ullah, W., Iqbal, J., Rahim, A., Alkahtani, H. K., Alghamdi, W., & Almagrabi, A. O. (2023). A new deep boosted CNN and ensemble learning based iot malware detection. *Computers & Security, 133*, 103385.

Khanzadi, M., Nasirzadeh, F., & Dashti, M. S. (2017). Fuzzy cognitive map approach to analyze causes of change orders in construction projects. *Journal of Construction Engineering and Management, 144*, 04017111.

Khazaeiathar, M., Hadizadeh, R., Fathollahzadeh, A. N., & Schmalz, B. (2022). Daily streamflow time series modeling by using a periodic autoregressive model (ARMA) based on fuzzy clustering. *Water, 14*, 3932.

Kikon, A., Dodamani, B. M., Barma, S. D., & Naganna, S. R. (2023). ANFIS-based soft computing models for forecasting effective drought index over an arid region of India. *AQUA—Water Infrastructure, Ecosystems and Society, 72*, 930–946.

Kilincer, I. F., Ertam, F., & Sengur, A. (2022). A Comprehensive intrusion detection framework using boosting algorithms. *Computers and Electrical Engineering, 100*, 107869.

Koşar, E., & Barshan, B. (2023). A New CNN-LSTM architecture for activity recognition employing wearable motion sensor data: Enabling diverse feature extraction. *Engineering Applications of Artificial Intelligence, 124*, 106529.

Kourehpaz, P., & Hutt, C. M. (2022). Machine learning for enhanced regional seismic risk assessments. *Journal of Structural Engineering, 148*, 04022126.

Kumar, S., Mallik, A., Kumar, A., Ser, J. D., & Yang, G. (2023). Fuzz-ClustNet: Coupled fuzzy clustering and deep neural networks for arrhythmia detection from ECG signals. *Computers in Biology and Medicine, 153*, 106511.

Latif, S. D., & Ahmed, A. N. (2023). Streamflow prediction utilizing deep learning and machine learning algorithms for sustainable water supply management. *Water Resources Management, 37*, 3227–3241.

Leite, B., da Costa, A. O. S., & da Costa, J. F. (2023). Multiobjective optimization of adiabatic styrene reactors using generalized differential evolution 3 (GDE3). *Chemical Engineering Science, 265*, 118196.

Li, P., & Abdel-Aty, M. (2022). Real-time crash likelihood prediction using temporal attention–based deep learning and trajectory fusion. *Journal of Transportation Engineering. Part A: Systems, 148*, 04022043.

Li, W., Huyan, J., Gao, R., Hao, X., Hu, Y., & Zhang, Y. (2021). Unsupervised deep learning for road crack classification by fusing convolutional neural network and k-means clustering. *Journal of Transportation Engineering. Part B: Pavements, 147*, 04021066.

Li, Q., Wang, K. P., Eacker, M., & Zhongjie, Z. (2016). Clustering methods for truck traffic characterization in pavement ME design. *ASCE-ASME Journal of Risk and Uncertainty in Engineering Systems. Part A: Civil Engineering, 3*, F4016003.

Liu, K., Zhang, R., Zhang, S., Chang, Y., Li, M., Wang, Y., Liu, Q., & Yu, D. (2023). Composition design of high-performance copper alloy by coupling artificial neural network and genetic algorithm. *Computational Materials Science, 229*, 112449.

Liu, X., Zhang, Y., Wang, J., Huang, H., & Yin, H. (2022). Multi-source(s) and multivariate ozone prediction based on fuzzy cognitive maps and evidential reasoning theory. *Applied Soft Computing, 119*, 108600.

Liu, Y., Liu, C., & Luo, X. (2021). Spatiotemporal deep-learning networks for shared-parking demand prediction. *Journal of Transportation Engineering. Part A: Systems, 147*, 04021026.

Luo, L., Wu, X., Hong, J., & Wu, G. (2022). Fuzzy cognitive map-enabled approach for investigating the relationship between influencing factors and prefabricated building cost considering dynamic interactions. *Journal of Construction Engineering and Management, 148*, 04022081.

Ma, J., Wang, J., Han, Y., Dong, S., Yin, L., & Xiao, Y. (2023). Towards data-driven modeling for complex contact phenomena via self-optimized artificial neural network methodology. *Mechanism and Machine Theory, 182*, 105223.

Madani, H., Kooshafar, M., & Emadi, M. (2020). Compressive strength prediction of nanosilica-incorporated cement mixtures using adaptive neuro-fuzzy inference system and artificial neural network models. *Practice Periodical on Structural Design and Construction, 25*, 04020021.

Madhuri, R., Sistla, S., & Raju, K. S. (2021). Application of machine learning algorithms for flood susceptibility assessment and risk management. *Journal of Water and Climate Change, 12*, 2608–2623.

Mangalathu, S., Karthikeyan, K., Feng, D. C., & Jeon, J. S. (2022). Machine-learning interpretability techniques for seismic performance assessment of infrastructure systems. *Engineering Structures, 250*, 112883.

Martinez-Gil, J., & Chaves-Gonzalez, J. M. (2022). Interpretable ontology meta-matching in the biomedical domain using mamdani fuzzy inference. *Expert Systems with Applications, 188*, 116025.

Mehryar, S., & Surminski, S. (2022). Investigating flood resilience perceptions and supporting collective decision-making through fuzzy cognitive mapping. *Science of the Total Environment, 837*, 155854.

Menaka, N., & Samraj, J. (2023). A hybrid convolutional neural network with long short-term memory (HCNN-LSTM) model based edge system recommendation (ESR) for cloud service providers. *Measurement: Sensors*, 29, 100886.

Mirzaie, N., Banihabib, M. E., Shahdany, S. M. H., & Randhir, T. O. (2021). Fuzzy particle swarm optimization for conjunctive use of groundwater and reclaimed wastewater under uncertainty. *Agricultural Water Management, 256*, 107116.

Mohammed, M. A. A., Khleel, N. A. A., Szabó, N. P., & Szucs, P. (2023). Modeling of groundwater quality index by using artificial intelligence algorithms in Northern Khartoum State Sudan. *Modeling Earth Systems and Environment, 9*, 2501–2516.

Morone, P., Yilan, G., & Imbert, E. (2021). Using fuzzy cognitive maps to identify better policy strategies to valorize organic waste flows: An Italian case study. *Journal of Cleaner Production, 319*, 128722.

Morris, C., Yang, J. J., Chorzepa, M. G., Kim, S. S., & Durham, S. A. (2022). Self-supervised deep learning framework for anomaly detection in traffic data. *Journal of Transportation Engineering. Part A: Systems, 148*, 04022020.

Muruganandam, S., Joshi, R., Suresh, P., Balakrishna, N., Kishore, K. H., & Manikanthan, S. V. (2023). A deep learning based feed forward artificial neural network to predict the k-barriers for intrusion detection using a wireless sensor network. *Measurement: Sensors, 25*, 100613.

Narayanan, S. L., Kasiselvanathan, M., Gurumoorthy, K. B., & Kiruthika, V. (2023). Particle swarm optimization based artificial neural network (PSO-ANN) model for effective K-barrier count intrusion detection system in WSN. *Measurement: Sensors, 29*, 100875.

Nayak, J., Naik, B., Dash, P. B., Souri, A., & Shanmuganathan, V. (2021). Hyper-parameter tuned light gradient boosting machine using memetic firefly algorithm for hand gesture recognition. *Applied Soft Computing, 107*, 107478.

Nogueira, S. C., de L,, Och, S. H., Moura, L. M., Domingues, E., Coelho, L. dos S., & Mariani, V. C. (2023). Prediction of the NOx and CO_2 emissions from an experimental dual fuel engine using optimized random forest combined with feature engineering. *Energy*, 280,128066.

Oqaidi, K., Aouhassi, S., & Mansouri, K. (2022). A comparison between using fuzzy cognitive mapping and machine learning to predict students' performance in higher education. 2022 *IEEE 3rd International Conference on Electronics, Control, Optimization and Computer Science* (ICECOCS), Fez, Morocco, 1–5.

Ozelim, L. C. d. S. M., Borges, L. P. d. F., Cavalcante, A. L. B., Albuquerque, E. A. C., Diniz, M. d. S., Góis, M. S., ... & Aquino, F. R. d. (2022). Structural health monitoring of dams based on acoustic monitoring, deep neural networks Fuzzy Logic and a CUSUM Control Algorithm. Sensors, 22, 2482.

Park, H. J., Kim, Y., & Kim, H. Y. (2022). Stock market forecasting using a multi-task approach integrating long short-term memory and the random forest framework. *Applied Soft Computing, 114*, 108106.

Pereira, I. P. C., Ferreira, F. A. F., Pereira, L. F., Govindan, K., Meidutė-Kavaliauskienė, I., & Correia, R. J. C. (2020). A fuzzy cognitive mapping-system dynamics approach to energy-change impacts on the sustainability of small and medium-sized enterprises. *Journal of Cleaner Production, 256*, 120154.

Perera, R., Torres, L., Ruiz, A., Barris, C., & Baena, M. (2019). An EMI-based clustering for structural health monitoring of nsm frp strengthening systems. *Sensors, 19*, 3775.

Pessoa, F. C. L., Blanco, C. J. C., & Gomes, E. P. (2018). Delineation of homogeneous regions for streamflow via fuzzy C-means in the amazon. *Water Practice and Technology, 13*, 210–218.

Prieto, A. J., Guiñez, F., Ortiz, M., & González, M. (2022). Fuzzy inference system for predicting functional service life of concrete pavements in airports. *Infrastructures, 7*, 162.

Prieto, A. J., Macías-Bernal, J. M., María-José, C., & Alejandre, F. J. (2017). Fuzzy modeling of the functional service life of architectural heritage buildings. *Journal of Performance of Constructed Facilities, 31*, 04017041.

Qian, L., Chen, Z., Huang, Y., & Stanford, R. J. (2023). Employing categorical boosting (Catboost) and meta-heuristic algorithms for predicting the urban gas consumption. *Urban Climate, 51*, 101647.

Qin, S., Xu, T., & Zhou, W. H. (2021). Predicting pore-water pressure in front of a TBM using a deep learning approach. *International Journal of Geomechanics, 21*, 04021140.

Rajeshkumar, G., Braveen, M., Venkatesh, R., Shermila, P. J., Prabu, B. G., Veerasamy, B., Bharathi, B., & Jeyam, A. (2023). Smart office automation via faster R-CNN based face recognition and internet of things. *Measurement: Sensors, 27*, 100719.

Ranawat, N. S., Prakash, J., Miglani, A., & Kankar, P. K. (2023). Performance evaluation of LSTM and Bi-LSTM using non-convolutional features for blockage detection in centrifugal pump. *Engineering Applications of Artificial Intelligence, 122*, 106092.

Sasan, G., Zahra, Z., Osama, M., & Sabah, A. (2019). Application of artificial neural network(s) in predicting formwork labour productivity. *Artificial Intelligence Applications in Civil Engineering, 2019*, 5972620.

Seaton, F. M., Barrett, G., Burden, A., Creer, S., Fitos, E., Garbutt, A., Griffiths, R. I., Henrys, P., Jones, D. L., Keenan, P., Keith, A., Lebron, I., Maskell, L., Pereira, M. G., Reinsch, S., Smart, S. M., Williams, B., Emmett, B. A., & Robinson, D. A. (2021). Soil health cluster analysis based on national monitoring of soil indicators. *European Journal of Soil Science, 72*, 2414–2429.

Shelia, V., & Hoogenboom, G. (2020). A new approach to clustering soil profile data using the modified distance matrix. *Computers and Electronics in Agriculture, 176*, 105631.

Solana-Gutiérrez, J., Rincón, G., Alonso, C., & García-de-Jalón, D. (2017). Using fuzzy cognitive maps for predicting river management responses: A case study of the esla river basin. *Spain, Ecological Modelling, 360*, 260–269.

Song, J., Kim, T., & You, D. (2023). Particle swarm optimization of a wind farm layout with active control of turbine yaws. *Renewable Energy, 206*, 738–747.

Sujith, M., & Jeon, J. S. (2019). Machine learning-based failure mode recognition of circular reinforced concrete bridge columns: Comparative study. *Journal of Structural Engineering, 145*, 04019104.

Sundar, L. S., & Mewada, H. K. (2023). Experimental entropy generation, exergy efficiency and thermal performance factor of Cofe2o4/water nanofluids in a tube predicted with ANFIS and MLP models. *International Journal of Thermal Sciences, 190*, 108328.

Sweidan, S., Zamzami, N., & Sabbeh, S. F. (2023). Fuzzy ontology-based approach for liver fibrosis diagnosis. *Journal of King Saud University—Computer and Information Sciences, 35*, 101720.

Takahashi, K., Sun, Z., Solé-Casals, J., Cichocki, A., Phan, A. H., Zhao, Q., Zhao, H.-H., Deng, S., & Micheletto, R. (2022). Data augmentation for convolutional LSTM based brain computer interface system. *Applied Soft Computing, 122*, 108811.

Tan, J., Wang, Q., Yan, K., Wei, X., & Fu, X. (2023). Saca-FI: A microarchitecture-level fault injection framework for reliability analysis of systolic array based CNN accelerator. *Future Generation Computer Systems, 147*, 251–264.

Taylan, O., Alkabaa, A. S., Alamoudi, M., Basahel, A., Balubaid, M., Andejany, M., & Alidrisi, H. (2021). Air quality modeling for sustainable clean environment using ANFIS and machine learning approaches. *Atmosphere, 12*, 713.

Tchupo, D. E., & Macht, G. A. (2022). Comparing fuzzy cognitive maps: Methods and their applications in team communication. *International Journal of Industrial Ergonomics, 92*, 103344.

Theisen, M. F., Flores, K. N., Balhorn, L. S., & Schweidtmann, A. M. (2023). Digitization of chemical process flow diagrams using deep convolutional neural networks. *Digital Chemical Engineering, 6*, 100072.

Tian, H., Yang, L., & Ju, B. (2023). Spatial correlation and temporal attention-based LSTM for remaining useful life prediction of turbofan engine. *Measurement, 214*, 112816.

Tirupathi, C., & Shashidhar, T. (2019). Fuzzy-based regional water quality index for surface water quality assessment. *Journal of Hazardous, Toxic, and Radioactive Waste, 23*, 04019010.

Tung, C. C., Lai, Y. Y., Chen, Y. Z., Lin, C. C., & Chen, P. Y. (2023). Optimization of mechanical properties of bio-inspired voronoi structures by genetic algorithm. *Journal of Materials Research and Technology, 26*, 3813–3829.

Tyagi, S., & Jha, V. (2023). Energy centric reputation index and fuzzy-based clustering for wireless sensor networks. *Applied Soft Computing, 146*, 110602.

Vasan, A., Raju, K. S., & Pankaj, S. (2021). Fuzzy optimization based water distribution network design using self-adaptive cuckoo search algorithm. *Water Supply, 22*, 3178–3194.

Vetrimani, E., Arulselvi, M., & Ramesh, G. (2023). Building convolutional neural network parameters using genetic algorithm for the croup cough classification problem. *Measurement: Sensors, 27*, 100717.

Vijayalakshmi, S., Jayashree, S., Elpiniki, P., & Suji, M. (2017). Application of fuzzy cognitive maps for crack categorization in columns of reinforced concrete structures. *Neural Computing and Applications, 28*, 107–117.

Vogeti, R. K., Bhavesh, R. M., & Raju, K. S. (2022). Machine learning algorithms for stream flow forecasting over lower godavari basin. *H2Open Journal, 5*, 670–685.

Wan, A., Chang, Q., AL-Bukhaiti, K., & He, J. (2023). Short-term power load forecasting for combined heat and power using CNN-LSTM enhanced by attention mechanism. *Energy, 282*, 128274.

Wang, D., Zhang, Z., Zhou, J., Zhang, B., & Li, M. (2022a). Comparison and analysis of several clustering algorithms for pavement crack segmentation guided by computational intelligence. *Computer Intelligence Neuroscience*, 8965842.

Wang, W., Lin, W., Wen, Y., Lai, X., Peng, P., Zhang, Y., & Li, K. (2023a). An interpretable intuitionistic fuzzy inference model for stock prediction. *Expert Systems with Applications, 213* (Part A), 118908.

Wang, X., Guo, G., Liu, S., Wu, Y., Xu, X., & Smith, K. (2020). Burst detection in district metering areas using deep learning method. *Journal of Water Resources Planning and Management, 146*, 04020031.

Wang, Z., Li, J., Rangaiah, G. P., & Wu, Z. (2022b). Machine learning aided multiobjective optimization and multi-criteria decision making: Framework and two applications in chemical engineering. *Computers & Chemical Engineering, 165*, 107945.

Xiang, Y., Wang, Z., Zhang, S., Jiang, L., Lin, Y., & Tan, J. (2024). Cross-sectional performance prediction of metal tubes bending with tangential variable boosting based on parameters-weight-adaptive CNN. *Expert Systems with Applications, 237* (Part A), 121465.

Xu, H., Wu, L., Xiong, S., Li, W., Garg, A., & Gao, L. (2023). An improved CNN-LSTM model-based state-of-health estimation approach for lithium-ion batteries. *Energy, 276*, 127585.

Xu, K., & Sun, Z. (2023). Predicting academic performance associated with physical fitness of primary school students using machine learning methods. *Complementary Therapies in Clinical Practice, 51*, 101736.

Xue, M., Zhou, L., Kojima, N., dos Muchangos, L. S., Machimura, T., & Tokai, A. (2018). Application of fuzzy C-means clustering to PRTR chemicals uncovering their release and toxicity characteristics. *Science of the Total Environment, 622–623*, 861–868.

Yang, H., Chen, Z., Li, Y., Yao, L., Wang, G., Deng, Q., Fu, P., & Wang, S. (2023). Modeling and optimization of graphene oxide (GO) membranes for nanofiltration with artificial neural networks. *Journal of Water Process Engineering, 55*, 104088.

Yuan, Z., Niu, M.-Q., Ma, H., Gao, T., Zang, J., Zhang, Y., & Chen, L. Q. (2023). Predicting mechanical behaviors of rubber materials with artificial neural networks. *International Journal of Mechanical Sciences, 249*, 108265.

Zafari, P., & Ghaemi, A. (2023). Modeling and optimization of CO_2 capture into mixed MEA-PZ amine solutions using machine learning based on ANN and RSM models. *Results in Engineering, 19*, 101279.

Zare, S. G., Alipour, M., Hafezi, M., Stewart, R. A., & Rahman, A. (2022). Examining wind energy deployment pathways in complex macro-economic and political settings using a fuzzy cognitive map-based method. *Energy, 238* (Part A), 121673.

Zarei, M., Bayati, M. R., Ebrahimi-Nik, M., Rohani, A., & Hejazi, B. (2023). Modelling the removal efficiency of hydrogen sulfide from biogas in a biofilter using multiple linear regression and support vector machines. *Journal of Cleaner Production, 404*, 136965.

Zaresefat, M., Derakhshani, R., Nikpeyman, V., GhasemiNejad, A., & Raoof, A. (2023). Using artificial intelligence to identify suitable artificial groundwater recharge areas for the iranshahr basin. *Water, 15*, 1182.

Zhang, J., Zhang, M., Feng, Z., Ruifang, L. V., Lu, C., Dai, Y., & Dong, L. (2023b). Gated recurrent unit-enhanced deep convolutional neural network for real-time industrial process fault diagnosis. *Process Safety and Environmental Protection, 175*, 129–149.

Zhang, P., Yang, Y., & Yin, Z. Y. (2021a). BiLSTM-based soil-structure interface modeling. *International Journal of Geomechanics, 21*, 04021096.

Zhang, P., & Yin, Z. Y. (2021). A novel deep learning-based modelling strategy from image of particles to mechanical properties for granular materials with CNN and Bi-LSTM. *Computer Methods in Applied Mechanics and Engineering, 382*, 113858.

Zhang, X., Jin, L., Cui, C., & Sun, J. (2021b). A self-adaptive multiobjective dynamic differential evolution algorithm and its application in chemical engineering. *Applied Soft Computing, 106*, 107317.

Zhang, Z., Al-Bahrani, M., Ruhani, B., Ghalehsalimi, H. H., Ilghani, N. Z., Maleki, H., Ahmad, N., Nasajpour-Esfahani, N., & Toghraie, D. (2023c). Optimized ANFIS models based on grid partitioning, subtractive clustering, and fuzzy c-means to precise prediction of thermophysical properties of hybrid nanofluids. *Chemical Engineering Journal, 471*, 144362.

Zhao, L., Wang, Q., & Hwang, B. G. (2022). How to promote urban intelligent transportation: A fuzzy cognitive map study. *Frontiers in Neuroscience, 16*, 919914.

Zhou, C., Li, H., Yang, J., Yang, Q., Yang, L., He, S., & Yuan, X. (2023). Fuzzy regular least squares twin support vector machine and its application in fault diagnosis. *Expert Systems with Applications, 231*, 120804.

Zhu, J., & Wang, Y. (2021). Feature selection and deep learning for deterioration prediction of the bridges. *Journal of Performance of Constructed Facilities, 35*, 040210781.

Zhuang, J., Wang, Y., Peng, S., Zhang, S., & Liu, Y. (2023). Siegel distance-based fusion strategy and differential evolution algorithm for cooperative spectrum sensing. *Digital Signal Processing, 142*, 104215.

Appendix A
Representative AI Tools and Data Sources Related to AI

Table A.1 presents insight into representative AI tools. Table A.2 presents representative data sources that may be useful while working on AI algorithms. The reader will also find several other tools and datasets (other than those mentioned in Table A.1 and. A.2) from various sources that can be explored.

K. Srinivasa Raju and D. Nagesh Kumar, *Artificial Intelligence and Machine Learning Techniques in Engineering and Management*,
https://doi.org/10.1007/978-981-96-2621-2

Table A.1 Insight of representative AI tools

Name of AI tool	Webpage link	Salient remarks#
Content generation		
ChatGPT	https://chat.openai.com/	It is a chatbot that responds to user queries appropriately and creates content and other applications
Google Bard AI	https://bard.google.com/u/1/chat https://bard.google.com/u/1/faq	It is a chatbot that responds to user queries appropriately and assists in translating languages, creating content, and interacting with other Google applications
Open AI Playground	https://platform.openai.com/playground https://gpt3demo.com/apps/openai-gpt-3-playground	It is a conversational chatbot that facilitates a wide range of tasks. It mainly focuses on technical research and development, allowing users to experiment with various ML algorithms and fine-tune them using custom datasets. It is highly beneficial in developing ML-based applications
The New Bing	https://www.microsoft.com/en-us/edge/features/the-new-bing?form=MT00D8	It is a chatbot that allows users to generate text, letters, and code. It also provides in-depth answers and summarizes information
Perplexity	https://www.perplexity.ai/	It is a chatbot and search engine that provides comprehensive solutions to user queries with the help of NLP and ML
Jasper chat	https://www.jasper.ai/chat	It facilitates a conversational environment, similar to a coworker or AI assistant. It does not require much knowledge to apply prompts efficiently. It allows many threads to explore different topic-related queries at a time
Chatsonic	https://writesonic.com/chat	It is a chatbot that can deliver conversational human-like responses to user queries and instantly help identify the frame of words to express ideas, develop information for advertising commercials, and get digital marketing plans
Claude	https://claude.ai/login?returnTo=%2F	It is trained on the most recent real-time data, enabling it to respond to current events
Llama	https://www.llama2.ai/	It is a chatbot that provides appropriate responses to user queries
Pi (Personal AI)	https://pi.ai/talk	It provides users access to high-quality conversation

(continued)

Table A.1 (continued)

Name of AI tool	Webpage link	Salient remarks[#]
Quora Poe	https://poe.com/	It utilizes advanced techniques in NLP and ML to quickly and effectively address user queries by exploring the extensive information repository accessible on Quora
DialoGPT	https://huggingface.co/docs/transformers/model_doc/dialogpt	It is trained with causal language modelling on conversational data. Influential at response generation in open-domain dialog systems
Character AI	https://beta.character.ai/	It is a chatbot that adopts natural language models to respond similarly to experts. Users can simultaneously design characters to interact with fictitious, historical, and celebrity personalities, gaining various perspectives
Replika	https://my.replika.com/signup/subscription	It is an interactive and personalized chatbot
Chai AI (Chai = chat + AI)	https://www.chai-ai.com/	It provides a text communication platform with AI chatbots. It is available on both Android and iPhone Operating System (iOS)
YouChat	https://you.com/search?q=Is+You.com+on+WhatsApp%3F&fromSearchBar=true&tbm=youchat&cid=c2_c2dc3872-c6f9-4119-a1d5-256902cf7aa0	It offers a prompt through which users can make a search query. As a result, the system provides the user with an AI-generated response and webpage links for verification Users can also explore images, videos, news articles, maps, and other relevant content
Copy AI	https://app.copy.ai/projects/34254106?tool=chat&tab=results https://www.elegantthemes.com/blog/marketing/copy-ai	It is a writing tool that uses ML to produce various forms of content, such as blog headlines, emails, social media material, and website copy
Frase	https://www.frase.io/tools/ai-content-generator/	It is a content generation tool that allows users to research, write, and optimize the content quickly and effectively
Fireflies.ai	https://fireflies.ai	It is a generative AI that uses ChatGPT to schedule meetings. It also generates transcripts and summaries for Zoom, Google Meet, and Microsoft Teams

(continued)

Table A.1 (continued)

Name of AI tool	Webpage link	Salient remarks#
Summarization		
Elicit	https://elicit.org/	It summarizes papers, extracts data, and synthesizes findings of the research papers to make an effective literature survey
PopAI	https://www.popai.pro/?utm_source=google&utm_medium=YM_popai&utm_campaign=0801&utm_term=in&utm_content=general_others&gclid=Cj0KCQjwmvSoBhDOARIsAK6aV7iAGfhMYOg56odIDXaPYT46jlaKYOtXzs4K6KWHQJZsiSTSHl2-EIcaAhymEALw_wcB	It helps prepare presentations, flow charts, coding answers, prompt generators, educational and professional writing
Learnt.ai	https://learnt.ai/	It majorly helps in content development by using appropriate prompts. Very useful for education purposes
Socrat.ai	https://socrat.ai/	It utilizes Google's AI and search technologies to facilitate the connection between students and educational resources available on the Internet
Coding related		
GitHub Copilot X	https://github.com/enterprise/trial?ref_cta=free%2520trial&ref_loc=banner&ref_page=blog	It assists users with various coding-related tasks and helps fix bugs in code
Amazon Codewhisperer	https://aws.amazon.com/codewhisperer/resources/#Getting_started/	It helps users overcome code-relevant errors and also provides valuable solutions
Paraphrasing		
Wordtune	https://app.wordtune.com/editor/documents/1e3272fc-b967-4e75-a4b4-45167f926d3d	It aids in rephrasing text, generating citations, summarizing long paragraphs into key sentences, checking grammar, and detecting plagiarism
Prepostseo	https://www.prepostseo.com/	It is a web-based program that allows content optimization, plagiarism detection, Paraphrasing, etc.
Quillbot	https://quillbot.com/settings?menu=plan	Paraphrasing tool, Grammatical checker
Spinbot	https://spinbot.com/	Paraphrasing tool, Grammatical checker

(continued)

Table A.1 (continued)

Name of AI tool	Webpage link	Salient remarks#
Word AI	https://wordai.com/	Paraphrasing tool
Speedwrite	https://speedwrite.com/creative-app	Paraphrasing tool
Anyword	https://anyword.com/paraphrasing-tool/	Paraphrasing tool
Hypotenuse AI	https://www.hypotenuse.ai/	Paraphrasing tool
Edit pad Paraphrasing Tool	https://www.editpad.org/tool/paraphrasing-tool	Paraphrasing tool
Good Content	https://www.semrush.com/goodcontent/paraphrasing-tool/	Paraphrasing tool
ProWritingAid	https://prowritingaid.com/paraphrasing-tool	Paraphrasing tool
SpinnerChief	https://www.spinnerchief.com/	Paraphrasing tool
DupliChecker	https://www.duplichecker.com/article-rewriter.php	Paraphrasing tool
Elsa speak	https://elsaspeak.com/en/inf	It helps to improve pronunciation and English-speaking skills
Virtual Assistant##		
Murf AI	https://murf.ai	It helps as a voice generator, eliminating the entire process of manually generating voiceovers and enabling users to produce human-like responses quickly
Siri	https://www.apple.com/in/siri/	It helps to send texts and make calls
Cortana	https://www.microsoft.com › en-us › cortana	It is a personal productivity assistant in Microsoft 365 and helps users achieve more outcomes with less effort
Cleo	https://web.meetcleo.com	It facilitates financial-related budgeting, saving, and building credit
Amazon Alexa	https://alexa.amazon.com	It offers customers natural voice experiences in a more intuitive way
Google Assistant	https://assistant.google.com	It helps to handle the schedules, set reminders, manage the schedule, manage the smart home, and do many more things

Remarks were taken from the relevant sources to convey the functionalities better

Many AI-based personal assistants (virtual assistants) work on mechanisms such as NLP to take text and voice commands. They can handle many activities similar to humans. For example, making telephone calls, scheduling meetings, creating text messages, setting reminders, calling taxis, managing workflow, reading the text, etc. These are expected to automate monotonous and time-consuming activities, making humans focus on essential tasks

Table A.2 Representative data sources

Data source	Link	Brief remarks
Kaggle	https://www.kaggle.com/datasets	263,555 free public datasets (as per the website) that can be used in AI and allied fields, namely, computer science, education, classification, computer vision, data visualization, pre-trained models, and many more Representative datasets include customer shopping trends, consumer behaviour and shopping habits, credit card fraud detection, global food price inflation, and heart attack risk prediction
Data Flair	https://data-flair.training/blogs/machine-learning-datasets/	70 + ML Datasets are part of this website. Some of the related datasets are Mall Customers, Iris, MNIST, The Boston Housing, Fake News Detection, SOCR data–Heights and Weights, Parkinson, Titanic, Uber Pickups, Chars74k, Credit Card Fraud Detection, Chatbot Intents, AI-generated Faces, and many more
DataONE	https://www.dataone.org/	It is a community-driven program giving avenues to data across multiple member repositories in the domain of earth and environment
DataPortals	https://dataportals.org	Detailed list of open data portals
Datasetlist.com	https://www.datasetlist.com	A list of ML datasets is available
Macgence	https://macgence.com/	It can design a customized AI-based data collection program that addresses the needs of various applications
Microsoft datasets	https://www.microsoft.com/en-us/research/tools/?	An index of datasets, software development kits, and other open-source codes created by Microsoft researchers is available. In addition, the website maintains a collection highlighting some of the tools

(continued)

Table A.2 (continued)

Data source	Link	Brief remarks
Mendeley Data	https://data.mendeley.com/	It is a free and cloud-based repository. Data can be stored, shared, and accessed anywhere
Dept. Mechanical Engg., Faculty of Engineering and Design, University Bath,	https://researchportal.bath.ac.uk/en/organisations/department-of-mechanical-engineering/datasets/	Website provided datasets related to various domains
Fraunhofer IPT and the Fraunhofer FFB	https://www.bigdata-ai.fraunhofer.de/s/datasets/index.html	ML datasets from the production environment compiled by the Fraunhofer: IPT (https://www.ipt.fraunhofer.de/) and the Fraunhofer FFB (https://www.ffb.fraunhofer.de/)
V7 labs	https://www.v7labs.com/blog/best-free-datasets-for-machine-learning	65 + free datasets for ML. Representative datasets include Open Dataset Aggregators, Public Government, Finance and Economics, Images for Computer Vision, Natural Language Processing, Data Visualization, Audio Speech, and Music
MIT Libraries	https://libguides.mit.edu/eecs/mldata	Electrical Engineering & Computer Science: Data sources for AI and ML
Electrical and Computer Engg., Grainger College of Engg., University of Illinois Urbana-Champaign	https://experts.illinois.edu/en/organisations/electrical-and-computer-engineering/datasets/	Website provided datasets related to various domains
Electronic Engineering, University of York, Faculty of Sciences	https://pure.york.ac.uk/portal/en/organisations/electronic-engineering/datasets/	Website provided datasets related to various domains
Electronic and Electrical Engg., Faculty of Engg., University of Strathclyde, Glasgow, UK	https://pureportal.strath.ac.uk/en/organisations/electronic-and-electrical-engineering/datasets/	Website provided datasets related to various domains
iguazio	https://www.iguazio.com/blog/best-13-free-financial-datasets-for-machine-learning/	The website provided 13 free financial datasets for ML
iMerit	https://imerit.net/blog/20-best-finance-economic-datasets-for-machine-learning-all-pbm/	The website provided 20 financial and economic datasets for ML
data.world	https://data.world/datasets/finance	The website provided 1096 finance-related datasets

Table A.2 (continued)

Appendix B
Representative Books and Journals on AI

Representative Books on AI

Alpaydin E (2016) Machine Learning: The New AI, The MIT Press.

Bhattacharyya S, Bhattacharyya S (2021) Practical Handbook of Machine Learning, GK Publications (P) Ltd.

Boden MA (2018) Artificial Intelligence: A Very Short Introduction, Oxford University Press.

Bonaccorso G (2018) Machine Learning Algorithms: Popular algorithms for data science and machine learning, Packt Publishing Limited.

Charniak E (2019) Introduction to Deep Learning, The MIT Press.

Chollet F (2017) Deep Learning with Python, Manning.

Courville A, Goodfellow I, Bengio Y (2016) Deep Learning, MIT Press.

de Prado ML (2018) Advances in Financial Machine Learning, Wiley.

Deisenroth MP, Faisal AA, Ong CS (2020) Mathematics for Machine Learning, Cambridge University Press.

Dutt S, Chandramouli S, Das AK (2018) Machine Learning, Pearson Education.

GÈron A (2022) Hands-On Machine Learning with Scikit-Learn, Keras, and TensorFlow: Concepts, Tools, and Techniques to Build Intelligent Systems, Shroff/ O'Reilly.

Grus G (2019) Data Science from Scratch: First Principles with Python, Shroff/ O'Reilly.

K. Srinivasa Raju and D. Nagesh Kumar, *Artificial Intelligence and Machine Learning Techniques in Engineering and Management*,
https://doi.org/10.1007/978-981-96-2621-2

Huyen C (2022) Designing Machine Learning Systems: An Iterative Process for Production-Ready Applications, Shroff/O'Reilly.

Jain A, Kapoor A, Fandango A (2018) Tensorflow Machine Learning Projects: Build 13 real-world projects with advanced numerical computations using the Python ecosystem, Packt Publishing Limited.

Kelleher JD (2019) Deep Learning, THE MIT Press.

Kelleher JD, Namee BM, Arcy A D' (2020) Fundamentals of Machine Learning for Predictive Data Analytics: Algorithms, Worked Examples, and Case Studies, The MIT Press.

Khemani D (2017) First Course in Artificial Intelligence, McGraw Hill Education.

Lapan M (2018) Deep Reinforcement Learning Hands-On: Apply modern RL methods, with deep Q-networks, value iteration, policy gradients, TRPO, AlphaGo Zero and more, Packt Publishing Limited.

Lee T, Singh VP, Cho KA (2021) Deep Learning for Hydrometerology and Environmental Science, Springer.

Liu Y (H) (2017) Python Machine Learning by Example: The easiest way to get into machine learning, Packt Publishing Limited.

Maheswari BU, Sujatha R (2021) Introduction to Data Science: Practical Approach with R and Python, Wiley.

Murphy KP (2012) Machine Learning: A Probabilistic Perspective, MIT Press.

Patterson DW (2015) Introduction to Artificial Intelligence and Expert Systems, Pearson.

Pradhan M, Kumar UD (2019) Machine Learning using Python, Wiley.

Qiufan C, Lee Kai-Fu (2021) AI 2041, WH Allen.

Rao RN (2022) Machine Learning in Data Science Using Python, Dreamtech Press.

Rose SL, Kumar LA, Renuka DK (2019) Deep Learning Using Python, Wiley.

Russell SJ, Norvig P (2022) Artificial Intelligence: A Modern Approach, Pearson.

Sutton RS, Barto AG, Bach F (2018) Reinforcement Learning: An Introduction, MIT Press.

Thareja R (2021) Data Science and Machine Learning with R, McGraw Hill.

Thareja R (2022) Data Science and Machine Learning using Python, McGraw Hill.

Representative Journals on AI

ACM Transactions on Intelligent Systems and Technology, Association for Computing Machinery, https://dl.acm.org/journal/tist

Advances in Data Analysis and Classification, Springer, https://www.springer.com/journal/11634

AI and Society, Springer, https://www.springer.com/journal/146

Applied Artificial Intelligence, Taylor and Francis, https://www.tandfonline.com/action/journalInformation?show=aimsScope&journalCode=uaai20

Applied Computational Intelligence and Soft Computing, Hindawi, https://www.hindawi.com/journals/acisc/

Applied Intelligence, Springer, https://www.springer.com/journal/10489

Artificial Intelligence, Elsevier, https://www.sciencedirect.com/journal/artificial-intelligence

Artificial Intelligence Review, Springer, https://www.springer.com/journal/10462

Applied Soft Computing, Elsevier, https://www.sciencedirect.com/journal/applied-soft-computing

Big Data and Cognitive Computing, MDPI, https://www.mdpi.com/journal/BDCC

Big Data Research, Elsevier, https://www.sciencedirect.com/journal/big-data-research

Cognitive Computation, Springer, https://www.springer.com/journal/12559

Cognitive Systems Research, Elsevier, https://www.sciencedirect.com/journal/cognitive-systems-research

Computational Intelligence, Wiley, https://onlinelibrary.wiley.com/journal/14678640?journalRedirectCheck=true

Computers and Education: Artificial Intelligence, Elsevier, https://www.sciencedirect.com/journal/computers-and-education-artificial-intelligence

Data and Knowledge Engineering, Elsevier, https://www.sciencedirect.com/journal/data-and-knowledge-engineering

Data Intelligence, The MIT Press, https://direct.mit.edu/dint

Data Mining and Knowledge Discovery, Springer, https://www.springer.com/journal/10618

Data Science and Engineering, Springer, https://www.springer.com/journal/41019

Discover Internet of Things, Springer, https://link.springer.com/journal/43926

Engineering Applications of Artificial Intelligence, Elsevier, https://www.sciencedirect.com/journal/swarm-and-evolutionary-computation

Expert Systems, Wiley, https://onlinelibrary.wiley.com/journal/14680394?journalRedirectCheck=true

Expert Systems with Applications, Elsevier, https://www.sciencedirect.com/journal/expert-systems-with-applications

Frontiers in Artificial Intelligence, Frontiers, https://www.frontiersin.org/journals/artificial-intelligence

Frontiers in Big Data, Frontiers, https://www.frontiersin.org/journals/big-data

Frontiers in the Internet of Things, Frontiers, https://www.frontiersin.org/journals/the-internet-of-things

Fuzzy Information and Engineering, Taylor and Francis, https://www.tandfonline.com/toc/tfie20/current

Fuzzy Optimization and Decision Making, Springer, https://www.springer.com/journal/10700

Fuzzy Sets and Systems, Elsevier, https://www.sciencedirect.com/journal/fuzzy-sets-and-systems

IEEE Big Data Mining and Analytics, IEEE, https://ieeexplore.ieee.org/xpl/aboutJournal.jsp?punumber=8254253

IEEE Computational Intelligence Magazine, IEEE, https://ieeexplore.ieee.org/xpl/aboutJournal.jsp?punumber=10207

IEEE Intelligent Systems, IEEE, https://ieeexplore.ieee.org/xpl/aboutJournal.jsp?punumber=9670

IEEE Internet of Things Journal, IEEE, https://ieeexplore.ieee.org/xpl/aboutJournal.jsp?punumber=6488907

IEEE Transactions on Artificial Intelligence, IEEE, https://cis.ieee.org/publications/ieee-transactions-on-artificial-intelligence

IEEE Transactions on Cognitive and Developmental Systems, IEEE, https://ieeexplore.ieee.org/xpl/aboutJournal.jsp?punumber=7274989

IEEE Transactions on Emerging Topics in Computational Intelligence, IEEE, https://ieeexplore.ieee.org/xpl/aboutJournal.jsp?punumber=7433297

IEEE Transactions on Evolutionary Computation, IEEE, https://ieeexplore.ieee.org/xpl/aboutJournal.jsp?punumber=4235

IEEE Transactions on Fuzzy Systems, https://ieeexplore.ieee.org/xpl/aboutJournal.jsp?punumber=91

IEEE Transactions on Information Theory, IEEE, https://ieeexplore.ieee.org/xpl/aboutJournal.jsp?punumber=18

IEEE Transactions on Knowledge and Data Engineering, IEEE, https://www.computer.org/csdl/journal/tk

IEEE Transactions on Neural Networks and Learning Systems, IEEE, https://ieeexplore.ieee.org/xpl/aboutJournal.jsp?punumber=5962385

IEEE Transactions on Pattern Analysis and Machine Intelligence, IEEE, https://www.computer.org/csdl/journal/tp

Information Fusion, Elsevier, https://www.sciencedirect.com/journal/information-fusion

Information Sciences, Elsevier, https://www.sciencedirect.com/journal/information-sciences

International Journal of Artificial Intelligence in Education, Springer, https://www.springer.com/journal/40593

International Journal on Artificial Intelligence Tools, World Scientific, https://www.worldscientific.com/worldscinet/ijait

International Journal of Big Data Intelligence, Inderscience Publishers, https://www.inderscience.com/jhome.php?jcode=ijbdi

International Journal of Data Science and Analytics, Springer, https://www.springer.com/journal/41060

International Journal of Fuzzy Systems, Springer, https://www.springer.com/journal/40815

International Journal of Intelligent Systems, Wiley, https://onlinelibrary.wiley.com/journal/1098111x?journalRedirectCheck=true

International Journal of Machine Learning and Cybernetics, Springer, https://www.springer.com/journal/13042

International Journal of Neural Systems, World Scientific, https://www.worldscientific.com/worldscinet/ijns

Internet of Things, Elsevier, https://www.sciencedirect.com/journal/internet-of-things

IoT, MDPI, https://www.mdpi.com/journal/IoT

Journal of Artificial Intelligence Research, Association for the Advancement of Artificial Intelligence, AI Access Foundation, https://www.jair.org/index.php/jair

Journal of Artificial Intelligence and Soft Computing Research, De Gruyter, https://sciendo.com/journal/jaiscr

Journal of Intelligent and Fuzzy Systems, IOS Press, https://www.iospress.com/catalog/journals/journal-of-intelligent-fuzzy-systems

Journal of Intelligent Systems, De Gruyter, https://www.degruyter.com/journal/key/jisys/html

Journal of Machine Learning Research, MIT Press, https://jmlr.csail.mit.edu/

Journal of Neural Engineering, IOP Publishing, https://iopscience.iop.org/journal/1741-2552

Knowledge and Information Systems, Springer, https://www.springer.com/journal/10115

Knowledge-Based Systems, Elsevier, https://www.sciencedirect.com/journal/knowledge-based-systems

Machine Learning, Springer, https://www.springer.com/journal/10994

Machine Learning and Knowledge Extraction, MDPI, https://www.mdpi.com/journal/make

Machine Learning: Science and Technology, IOP Publishing, https://iopscience.iop.org/journal/2632-2153

Nature Machine Intelligence, Natureportfolio, https://www.nature.com/natmachintell/

Neural Computation, The MIT Press, https://direct.mit.edu/neco

Neural Computing and Applications, Springer, https://www.springer.com/journal/521

Neural Networks, Elsevier, https://www.sciencedirect.com/journal/neural-networks

Neural Processing Letters, Springer, https://www.springer.com/journal/11063

Neuro Computing, Elsevier, https://www.sciencedirect.com/journal/neurocomputing

Neuro informatics, Springer, https://www.springer.com/journal/12021

Progress in Artificial Intelligence. Springer, https://www.springer.com/journal/13748

Soft Computing, Springer, https://www.springer.com/journal/500

Statistical Analysis and Data Mining, Wiley, https://onlinelibrary.wiley.com/journal/19321872

Statistics and Computing, Springer, https://www.springer.com/journal/11222

Swarm and Evolutionary Computation, Elsevier, https://www.sciencedirect.com/journal/swarm-and-evolutionary-computation

Swarm Intelligence, Springer, https://www.springer.com/journal/11721

Wiley Interdisciplinary Reviews: Data Mining and Knowledge Discovery, Wiley, https://wires.onlinelibrary.wiley.com/journal/19424795

Note Information may (or may not) be relevant to some of the readers. Note that we are not responsible for the website content and related inconvenience caused to the reader, if any, at later stages.

Index

K. Srinivasa Raju and D. Nagesh Kumar, *Artificial Intelligence and Machine Learning Techniques in Engineering and Management*,
https://doi.org/10.1007/978-981-96-2621-2

MIX
Papier aus verantwortungsvollen Quellen
Paper from responsible sources
FSC® C105338

If you have any concerns about our products,
you can contact us on
ProductSafety@springernature.com

In case Publisher is established outside the EU,
the EU authorized representative is:
Springer Nature Customer Service Center GmbH
Europaplatz 3, 69115 Heidelberg, Germany

Printed by Libri Plureos GmbH
in Hamburg, Germany